Reviews of

106 Physiology, Biochemistry and Pharmacology

Editors

P. F. Baker, London · H. Grunicke, Innsbruck
E. Habermann, Gießen · R. Jung, Freiburg
R. J. Linden, Leeds · P. A. Miescher, Genève
H. Neurath, Seattle · S. Numa, Kyoto
D. Pette, Konstanz · B. Sakmann, Göttingen
W. Singer, Frankfurt/M · U. Trendelenburg, Würzburg
K. J. Ullrich, Frankfurt/M

With 57 Figures

Springer-Verlag
Berlin Heidelberg GmbH

ISBN 978-3-662-31146-2 ISBN 978-3-540-47713-6 (eBook)
DOI 10.1007/978-3-540-47713-6

Library of Congress-Catalog-Card Number 74-3674

Originally published by Springer-Verlag Berlin Heidelberg New York in 1987
Softcover reprint of the hardcover 1st edition 1987

2127/3130-543210

Contents

Indexed in Current Contents

Rev. Physiol. Biochem. Pharmacol., Vol. 106

The Preparation of Visually Guided Saccades

BURKHART FISCHER

Contents

Abteilung Klinische Neurologie und Neurophysiologie, Universität Freiburg, Hansastr. 9, 7800 Freiburg, FRG

1 Introduction

The large amount of data on the initiation of saccadic eye movements that has been published in recent years makes a review both necessary and almost impossible at the same time. Psychophysical work in man, behavioral studies in animals, and neurophysiological and anatomical experiments in monkeys and cats must be considered as well as observations on neurologic and psychiatric patients. Since it is controlled by a large number of brain structures, including different cortical and subcortical visual areas, the saccadic system is far from operating like a simple reflex. Even though there are good reviews of parts of these structures, a synoptic view of how the parts of the whole system may act together during the initiation of a saccade is lacking.

This article reviews mainly the work published in the last 5 years from the point of view of a hypothetical conceptual frame work which – right or wrong – provides the reader with a context to comprehend the material.

1.1 Functions of the Saccadic System

Jung (1972) considered the saccadic system as a spatial location system which selects the next visual target, locates it by foveal vision, terminates the preceding fixation, and cancels its residual image. This description of many functions of saccades will be specified in terms of both physiology and anatomy.

When the stream of visual information originating in the retinal ganglion cells is passed through the lateral geniculate nucleus to the primary visual cortex (prestriate areas V2 to V5), there occurs in addition a wide distribution to many other brain structures: visual messages reach the pulvinar, the superior colliculus, the parietal cortex (Areas 5 and 7), the inferotemporal cortex (area 20), the frontal eye fields (area 8), the prefrontal cortex, the brain stem, and the cerebellum. In terms of function this wide anatomical distribution is not surprising, because the stream of visual information leads not only to visual perception of objects (e.g., their identification and their location in space) but also to body movements and behavior modulated by vision.

Vision and visually guided movements need the coordinating functions of higher processes like attention and memory. Therefore, the corresponding brain structures must have access to visual input and some control over the motor output.

In this article those physiological and psychological processes will be considered that are intermediate between afferent messages of vision and the efferent performance of the oculomotor system.

2 Seeing with Moving Eyes

Under natural viewing conditions the direction of gaze changes in rapid steps (saccades) from one point of interest to the next. These saccades occur about 2 to 3 times a second. They are usually below 15° in size (Frost and Poeppel 1976) and can have all directions between up and down and right and left. In the period between two consecutive saccades several events and processes must take place.

Determination of Coordinates. Saccades are goal-directed toward a preselected target in the visual surround. This requires precomputation of the target location (Becker and Juergens 1979). Where and when does this precomputation take place? Is the target position represented in Cartesian or cylinder coordinates (or others)? Is this system fixed in spatial or retinal coordinates?

Decision. Eye movements, although controlled by several reflex systems (e.g., vestibular ocular reflex), are also under voluntary control. Any voluntarily initiated saccade therefore needs a decision as to whether or not it should be executed. The problem of decision making arises in the context of any voluntary movement and in the context of fixation as well as of attention.

Attention. Usually, the direction of gaze and the direction of attention coincide. We will, however, discuss below their relation to fixation. One knows from introspection that the directions of gaze and attention can also be dissociated: the eye, for example, looks straight ahead, but attention can be simultaneously directed to the peripheral field. Which are the mechanisms of visual attention and how do they modulate the saccadic system on the one hand, and how do they interact with the mechanisms that provide stable fixation on the other? A shift of attention (similar to a shift of fixation) is under voluntary control and can take place without any external sensory event or any motor event. It requires, however, an awake and conscious subject.

Fixation. At the end of a saccade the fovea is centered on a preselected (see above) new target which is then fixated for some time. Fixation requires the cooperation of all extraocular muscles; the position of the eyeball in the orbit (not taking into account head and body movements) must be controlled with respect to the fixated target. Which are the (visual?) centers that provide the appropriate information to the oculomotor system? These centers must have access to foveal vision but must also be under the control of a nonvisual input which provides the information about whether or not the target presently projected onto the center of the fovea is to be fixated. The corresponding activity therefore is under voluntary control and can be modulated without an external sensory event, because one can stop fixation and change the direction of gaze at one's own will.

Stability. The saccade itself rapidly (velocity betwenn 20 °/s and 600 °/s depending on its size) sweeps the receptor surface across the retinal image. Yet one does not perceive a corresponding movement of the visual scene or recognize the new position of its image at the retina. Rather, one perceives a "stable world". How this is achieved by an appropriate interaction between vision and rapid eye movements is still an open question, which has been discussed in handbook reviews dating from Hering (1879) to MacKay (1973).

Memory. Saccades can be made toward spatial locations that the subject or monkey has to remember. The use of memory and the way in which its contents are fed into the oculomotor system are questions that can be tackled in both psychophysical and neurophysiological experiments.

All these processes (with the possible exception of the use of memory signals) must take place before the next saccade is executed, thereby initiating a new cycle.

3 Three Loops – Three Processes

To facilitate review of the different contributions to the problems set out above, the schematic drawing which constitutes Fig. 1 will be used as a conceptual frame work within which one may understand more easily the progress already made and the gaps left in our knowledge. Figure 1 illustrates that attention (A) is considered as a separate but closely related process which directly or indirectly has access to visual input (V) on the one hand and which feeds into the subsequent process of decision making (D) on the other hand. Visual input is also distributed to two other processes, and this distribution explains why visually guided saccades can be elicited through different pathways. The two other processes are

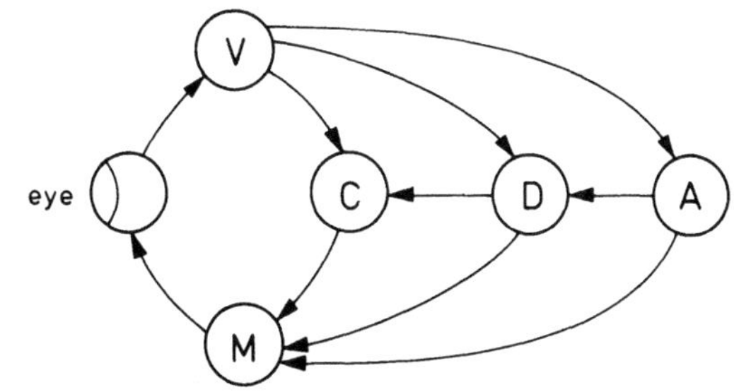

Fig. 1. Scheme of the three-loop concept of preparation of visually guided saccades. Visual information (*V*) from the retina of the eye is distributed to three different processes – *A, D,* and *C* – all of which send their output to the oculomotor center (*M*). *A,* attentional mechanisms; *D,* decision making; and *C,* computation of coordinates. The three loops are interconnected: *A* with *D* and *D* with *C*

decision making and computation (C) of saccade metrics. All three processes are also directly connected to the efferent oculomotor system (M) in the brain stem.

While Fig. 1 is mainly used to guide the reader through the material reviewed, it will be considered at the end to see whether or not it can be modified to describe the functional and anatomical aspects of the preparation of visually guided saccades as far as we understand them today.

The three loops and their interconnections shown in Fig. 1 have been derived mainly from reaction time studies in monkey and man. The basic observation is that the spectrum of the reaction times exhibits at least three different peaks (see below).

The idea behind saccadic reaction time studies is simply that all processes involved in the preparation of the eye movement, including afferent and efferent delay times, contribute to the reaction time. If one manages to exclude one or the other of these processes from the period of time between the external command to execute the eye movement and its actual beginning, the reaction time should be correspondingly shorter. Following up this idea has led to a number of new observations both in man and monkey, which are described below.

On the other hand, the three-loop concept rests on experiments utilizing electrical stimulation to elicit saccades and lesion experiments, which altogether show that the saccade system can be activated through different pathways.

4 First Loop – Computation of Saccade Metrics

The minimum requirement for a correct visually guided saccade is that its size and amplitude be available to the oculomotor centers. Since saccades are preprogrammed, the time for the computation must be included in the reaction time. From the work of Becker and Juergens (1979) it is known that this process is the last one in the preparation. In a double-step paradigm the saccade to the first target step (let us say to the right) could be cancelled and replaced by a saccade to the left if the second target step (to the left) occurred about 80 ms before the saccade. If one enables the system to complete all preparations except the computation of the target location, one would therefore expect extremely short reaction times when measured from the onset of the target.

4.1 Express Saccades

The typical value of monkey or human saccadic reaction time is in the order of 200 ms or a little less. These values are obtained when the central fixation point is turned off at the same time the peripheral target appears. Until recently, there was little doubt, if any, that a monkey could not react any faster than 150 ms.

In 1983 Fischer and Boch reported the existence of visually guided saccades that were characterized by their extremely short reaction times in the order of 70 ms. They called these saccades express saccades to

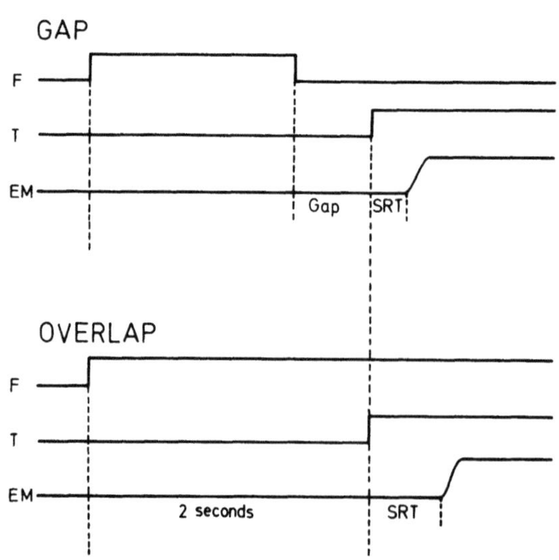

Fig. 2. Temporal sequence of the occurrence of the fixation point (*F*), the target (*T*), and the eye movement (*EM*). Saccadic reaction times (*SRT*) are measured from target onset. Note that in the gap task (*upper*) and in the overlap task (*lower*) the direction, size, and target for the saccade are the same

differentiate them from regular saccades with longer reaction times. The phenomenon of express saccades was first observed when the fixation point was extinguished about 200 ms (gap) before the peripheral target became visible (even if the target location was unpredictable). Under these conditions (outlined in Fig. 2) a clearly bimodal distribution of reaction times occurred, as can be seen from Figure 3A. The first narrow peak represents the population of express saccades and the second peak consists of the fast regular saccades. If a monkey had to make several saccades to a target on the same day, with the only difference that the fixation point remained visible (overlap paradigm, Fig. 2), the reaction times were broadly scattered and rather long, as can be seen from Fig. 3B. These saccades are called slow regular saccades.

Further experiments have shown that the spectrum of the saccadic reaction times of the monkey changes systematically with daily practice (Fischer et al. 1984). This phenomenon will be considered below.

Although it was supposed at the time of the first experiments that the introduction of a temporal gap between fixation point offset and target onset was a necessary condition for the occurrence of express saccades, it became clear later (Boch and Fischer 1986) that monkeys could execute express saccades even in the presence of a continuously visible fixation point (Fig. 3B).

The existence of express saccades in man was shown by Fischer and Ramsperger in 1984. Human subjects also produced a bimodal distribution of reaction times when the gap paradigm was used. The first peak,

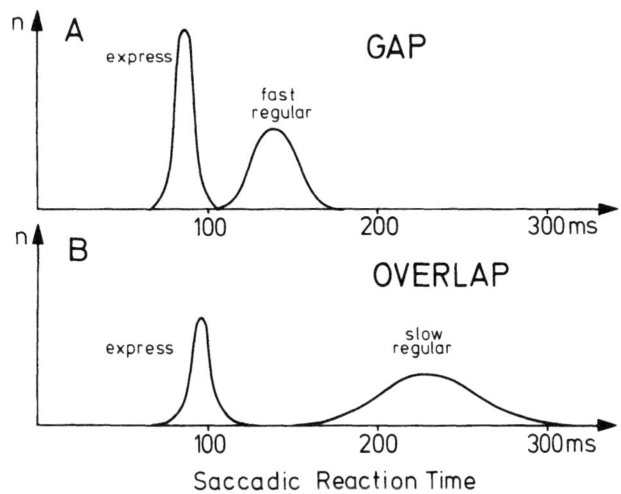

Fig. 3A, B. Schematic summary of saccadic reaction times obtained in the gap paradigm **(A)** and the overlap paradigm **(B)**

representing the express saccades, occurs at about 100 ms and in most subjects is clearly separated from a second peak at around 150 ms. Again, as in the monkey, the gap was not a necessary condition for the occurrence of express saccades, and the number of such saccades could be increased through daily practice by the subjects (Fischer and Ramsperger 1986).

Since the reaction time of the express saccades is so short, a reasonable part of it must be due to afferent delays in the retina and efferent delays between the oculomotor nuclei and the beginning of the actual movement of the eyeball. In fact, the reaction time of express saccades varies systematically with the physical nature of the target and target position at the retina (Boch et al. 1984): with increasing luminance of target (from threshold up to 1.5 log units and above) the reaction times of the express saccades in the monkey decrease from 120 ms to below 70 ms. With increasing target size the reaction times also decrease, reach a minimum value for a critical target size, and rise again. The critical target size increases with the eccentricity of the target, i.e., with the size of the corresponding saccade.

Subtracting the latency of retinal ganglion cell responses (Holz et al. 1982) and the latency of saccades elicited electrically (Robinson 1972; Schiller and Stryker 1972) from the shortest reaction time of express saccades leads to a central preparation time of only about 20 ms. The visual latencies of the prestriate cortex (Boch and Fischer 1983) and the frontal eye fields (Goldberg and Bushnell 1981) show that both cortical structures react to the onset of a stimulus in the cells' receptive fields after the express saccade has already been initiated. This immediately implies that the visual structures which can contribute to the initiation of express saccades can include only the primary visual cortex and the superior colliculus.

The further observation that express saccades are usually correct in direction and amplitude leads to the conclusion that the only process included in the reaction time of express saccades is the computation of target location at the retina. This conclusion is in agreement with that of Becker and Juergens (1979), who used double-step stimuli in man. The main result is that the preprogrammed saccade can be reprogrammed if the appropriate command is available 80 ms before the beginning of the eye movement. The experiment shows not only that saccades are preprogrammed but also that there is a last step in the preparation which has to do with the direction and size of the planned saccade. The express saccade, therefore, is an eye movement, the preparation of which has been completed with the sole exception of the final step, i.e., the computation of the (retinal) coordinates and the corresponding size and direction of the eye movement.

The overlap-gap paradigm has been used in other studies as well (Saslow 1967; Reulen 1984a; Reulen 1984b), but their authors did not report the existence of a separate peak in the distribution of the saccadic reaction times possibly corresponding to the express saccades. Reulen (1984a) speculated that the drastic decrease in saccadic reaction times, in the gap experiment – as compared to the overlap experiment – was a consequence of the offset of the fixation point. This interpretation is ruled out owing to the occurrence of extremely short latency saccades in the overlap paradigm, depending on the direction of visual attention with respect to the fixation point (see below).

4.2 Superior Colliculus

The shortest pathway from the retina to the oculomotor centers includes the superior colliculus. This already suggests that express saccades are mediated by the collicular pathway and that the colliculus is involved in the computation of saccade metrics. The direct evidence, however, comes from lesion experiments. Sandell et al. (1984) have shown that only express saccades, not regular saccades, are abolished when the superior colliculus is destroyed. On the other hand, express saccades can still be executed by monkeys without the frontal eye fields. In contrast to the multiple pathways used by regular saccades, there may exist only one pathway for express saccades.

4.3 Retinal Versus Spatial Coordinates

The target location measured primarily in retinal coordinates must be transformed into motor coordinates: the extraocular muscles must produce a saccade of appropriate size and direction. In principle, it is possible that the coordinate systems are both centered at the fovea, because any eye movement will move the sensory surface such that the next saccade can again be programmed in retinal coordinates. On the other hand, the sensory and motor frame of reference can be different such that the eye movement is programmed in a spatially centered frame, regardless of the position of the eye. Such a spatiotopic arrangement would have the advantage that it could be used to determine correct saccades to targets of other sensory modalities. To determine whether saccades are spatially or retinocentrically coded, monkeys trained to make saccades under different conditions have been used, while at the same time single-cell activity was recorded from brain structures that possibly mediate the interaction.

From the previous section it is already clear that the superior colliculus is a good candidate for such a structure. In fact, Mays and Sparks (1980b) and Sparks and Mays (1983) reported that monkeys could acquire target position correctly even if the eye was driven to another position in the orbit within the time of target appearance and the beginning of the saccade. This was done by electrical stimulation of the superior colliculus. Sparks and Porter (1983) then showed that eye movement related cells in the superior colliculus discharge before these compensatory saccades, and they concluded that the superior colliculus contains a spatial map of motor error. Moreover, it was demonstrated by Jay and Sparks (1984) that auditory receptive fields of superior colliculus cells shift with eye position, allowing the auditory and visual maps to remain in register. This again implies that the superior colliculus participates in the final process of computing target location and, more specifically, that the result of this computation is in a spatial, not a retinal, frame of coordinates. The question, however, of how and where this computation takes place is still open.

Mays and Sparks (1980a) showed that visual and saccade-related neural activity in the superior colliculus can be dissociated: saccade-related activity occurs irrespective of the retinal location stimulated to elicit the saccade. They concluded that "the superior colliculus must generate or receive a signal that combines retinal error and eye-position information". Following along those lines, Schiller and Sandell (1983) destroyed the colliculus in monkeys and found that these animals could still compensate for an electrically induced perturbation of the eyeball prior to a visually triggered saccade. This suggests that the computation takes place below the level of the colliculus, probably in the brain stem.

The combined information is also received by the frontal eye fields (Schiller and Sandell 1983), the postarcuate cortex (area 6) (Gentilucci et al. 1983), and the central thalamus (Schlag and Schlag-Rey 1984). The distribution of the combined information about the metrics of saccades in a body frame work makes sense because goal-directed eye movements must be in register with movements of the head and limbs, irrespective of the sensory system that delivers the spatial location of objects.

4.4 Striate Cortex

In order to obtain the spatial coordinates, two items of information must be available: (a) exact retinal target position and (b) exact eye position. The latter is not provided by proprioceptive feedback from the extraocular muscles but may stem from a corollary discharge originating in the oculomotor centres in the brain stem and is available at the superior colliculus (Richmond and Wurtz 1980; Guthrie et al. 1983). The pathway that provides the retinal target location seems to use the visual cortex, because punctate lesions (made chemically or surgically) result in a failure to acquire correct target position but do not abolish saccades altogether (Newsome et al. 1985). It seems that monkeys with striate cortical lesions do not see the target. In conclusion, seeing the target is a necessary condition for making correct saccades.

If this is the case and if computation of target coordinates in a motor frame work is the only central preparatory process included in the reaction time of express saccades, monkeys with striate cortical lesions should also be unable to execute express saccades. Some corresponding experiments in fact confirm this conclusion (Boch 1986).

Since electrical stimulation of the occipital cortex and the parietal cortex leads to saccadic eye movements (Wagman et al. 1958), at least if the animal is not fixating actively (see below), it is clear that these cortical structures contribute to the initiation of saccades. After lesions of the colliculus have occurred, electrical stimulation of these areas is no longer effective, a fact which suggests that the colliculus mediates these cortical contributions (Schiller 1977; Keating et al. 1983).

4.5 Stability of Vision

The result of the computation of saccade size and direction could in principle be used as a corollary message that informs the visual system about the occurrence of the eye movement in such a way that no movement is perceived during the saccade.

Several attempts have recently been made to demonstrate the existence of cells that differentiate between real and self-induced movements. They have been found in the superficial layers of the superior colliculus (Robinson and Wurtz 1976) and in the pulvinar (Robinson and Petersen 1985). The pulvinar, which receives projections from the colliculus as well as from the striate cortex (Campos-Ortega and Hayhow 1972), sends fibers to the extrastriate cortex. Recording in the prelunate cortex, Fischer et al. (1981) observed only a small percentage of cells that differentiated between the two movements. They reported that the overall response during self-induced movements was weaker, because most cells did not respond to stimulus velocities as fast as saccadic velocities.

Cells in the striate cortex mostly prefer slow stimulus movements (in the order of less than $10^{o}/s$) and those few cells that still respond to fast movements do not differentiate (Wurtz 1969; Fischer et al. 1981). Therefore, it has been suggested that the colliculus and the frontal eye fields are responsible for the differentiation. However, lesion experiments by Collin and Cowey (1980) have shown that neither of these two structures can mediate any kind of corollary discharge, because their absence would lead to confusion of real and self-induced movements.

4.6 Remembered Target Location

Target location is usually signaled directly by retinal information. But monkeys as well as man can remember where a target appeared and can execute correct saccades to the remembered target location. It is clear that the contents of the memory must finally be available at the level of the motor neurons. Little is known about where the memory is located and how it collaborates with the visual and motor mechanisms. We know that cells in the frontal eye fields discharge before saccades to target locations that the monkey has to remember (Bruce and Goldberg 1985), but this does not necessarily mean that the frontal eye fields have information about the location. It may rather be that the frontal eye fields discharge before any purposive saccade, as has been discussed by Goldberg and his coworkers.

Boch and Fischer (1986) trained monkeys to make saccades to a remembered target location. The animals had to fixate a central fixation spot and the target was flashed briefly (20 ms) while fixation was still maintained. Some time later the fixation spot was turned off and the animals were required to make a saccade to the previously illuminated location. The reaction time in this task was about 200 ms from fixation spot offset. If the target remained visible instead of being flashed, the reaction times were about the same. If, however, the target was turned on shortly after the offset of the fixation spot, the reaction times were considerably shorter. This result shows that the time taken to acquire target position from memory or from a continuously visible light spot is about the same. Target position can be acquired much faster if the visual system supplies the oculomotor system with an "on" response.

4.7 Corrective Saccades

Visually guided saccades sometimes miss the target, undershoots being more frequent than overshoots. The failure to reach the target at once means either that the computation of the metrics has not been completed or that its result came out wrong, or both.

We have argued above that the reaction time of express saccades includes only the time taken to complete this computation. If, therefore, a saccade is made after a latency, which is still shorter, one expects these "anticipatory" saccades to miss the target. This, in fact, has been reported by Fischer and his coworkers for both monkey and man. Interestingly, some of these anticipatory saccades were corrected after very short intersaccadic intervals, with the additional feature that the corrective saccades began at a constant time after target onset which is just as long as the reaction time of a correct express saccade. This is illustrated in Fig. 4A, which shows selected eye movement traces from two subjects performing the gap task. Figure 4B shows the distribution of the reaction time of the first (anticipatory) saccade and of the second (corrective) saccade, both measured from the offset of the fixation point. This observation indicates that the first saccade leaves a message about where the eye will be at its end. The second saccade, which is a corrective express saccade, can use this information and brings the fovea right on target.

Incorrect saccades also give an insight into the question of the differential programming of size and direction of the eye movement. Monkeys required to make saccades to target locations which were completely unpredictable often looked in the right direction but undershot the target. This implies that the computation of the direction and amplitude are different processes. Furthermore, if one randomizes the direction of the required saccade between right and left but keeps the distance constant, the reaction time of express saccades is increased by about 15 ms as compared to the situation in which the target location is always the same. If, on the other hand, one randomizes the distance but keeps the direction constant, such an increase in the reaction time is not observed.

Incorrect saccades can be elicited in laboratory conditions, in which subjects are presented with two simultaneously occurring targets. Coren and Hoenig (1972) and Findlay (1982) reported that under these conditions the saccade falls at about the "center of gravity" of the stimulus configuration (global effect). Obviously, the mechanism responsible for the computation of the metrics is influenced by both stimuli and there is little voluntary control by which to avoid the influence of the distracting target (Findlay 1982).

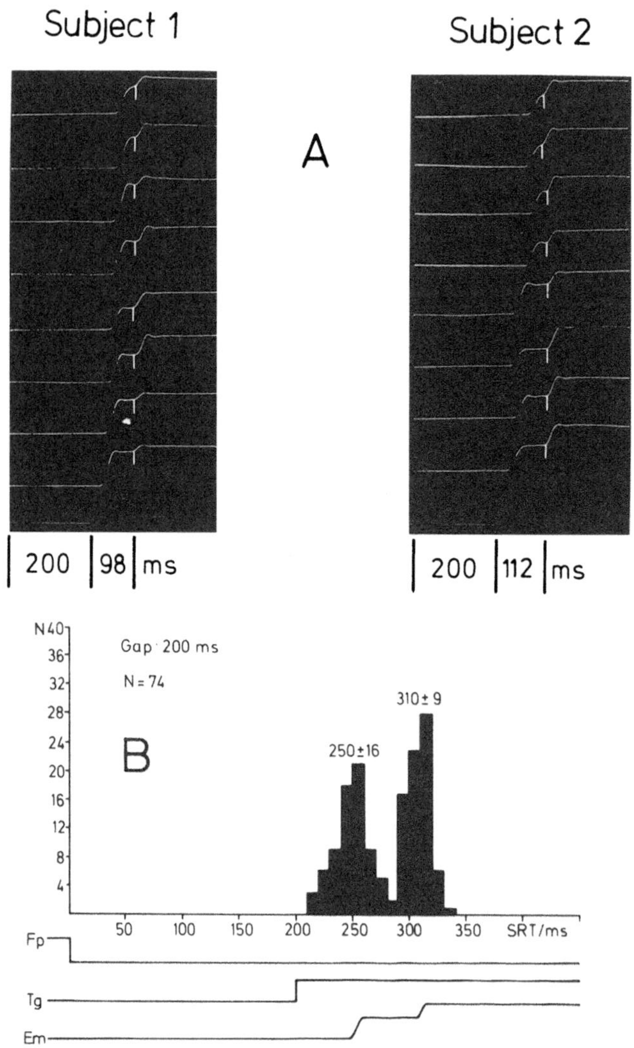

Fig. 4A, B. The phenomenon of corrective express saccades. **A** Eye movement traces from trials where the subject anticipates the target occurrence, misses the target position, and corrects at the time of an express saccade. **B** Distribution of reaction time of first (anticipatory) saccade and of second (corrective) saccade

Another experiment for exploring the mechanisms of the computation of saccade metrics is to shift the target after the primary saccade has been made. Deubel et al. (1982) have shown in man that the subsequent (secondary) saccade, which has to compensate for the externally induced error, can be modified only if the second target step occurs earlier than 60 ms after the end of the primary saccade. If the target was blanked at the beginning of the primary saccade and reilluminated at the shifted position only some time later, the corrective saccades had a constant latency of about 120 ms. This is quite close to the reaction time of the express saccades. In fact, the blanking of the target induces a gap within which the subject has no target to fixate or to pay attention to – a situation in which many express saccades can be obtained.

5 Second Loop – Decision Making

As already pointed out, there are at least two pathways that provide visual information to the saccade system, the first including the superior colliculus and striate cortex, and the second including the frontal eye fields (Schiller et al. 1979; Schiller et al. 1980). The main structures to consider, therefore, are the frontal eye fields and their connections to the first loop. Of course, the frontal eye fields are also connected to many other cortical and subcortical structures, such as the prefrontal cortex anteriorly and the premotor cortex posteriorly. These connections and their functional modulation of the saccadic system have been excluded from this review.

5.1 Frontal Eye Fields

The physiology of the frontal eye fields was recently reviewed by Bruce and Goldberg (1984). Here we consider only those aspects of recent work that might help to understand the role of the second loop in initiating saccades.

The frontal eye fields have access to visual information because neurons in this area respond to light stimuli (Wurtz and Mohler 1976b; Rizzolatti et al. 1981). They project to the oculomotor centers, bypassing the superior colliculus, because saccades can be executed even if the superior colliculus is destroyed.

According to the work of Goldberg there exists in the frontal eye fields a group of cells which are activated prior to visually guided saccades (Goldberg and Bushnell 1981). There are also cells which discharge before saccades to locations that the monkey has to remember, i.e., these cells do not necessarily need a visual input (Bruce and Goldberg 1984). However, they do not discharge prior to spontaneous saccades in the dark (Bizzi 1968). Microstimulation of the frontal eye fields elicits saccades of specific size and direction, which correspond to the movement fields of the cells recorded in that region (Bruce et al. 1985).

The conclusion drawn from these observations is that the frontal eye fields are active before purposive saccades that the monkey executes at will (Bruce and Goldberg 1985). Figure 1 therefore suggests that the second loop has to do with decision making.

Like the superior colliculus, the frontal eye fields receive information about eye position, because saccade-related movement cells shift their movement fields according to the last saccade (Goldberg and Bruce 1981). It has been suggested that those cells in the frontal eye fields that discharge after a saccadic eye movement receive their input from brain stem oculomotor structures to remap the reference frame (Bruce and Goldberg 1984).

Despite the great similarity of the physiological properties of neurons in the frontal eye fields and the superior colliculus, there are two qualitative differences: first, presaccadic activation in the colliculus occurs independently of whether or not the saccade is intentional; second, express saccades can be mediated by the collicular pathway alone, whereas this is not the case for the pathway through the frontal eye fields. From these data it seems that the two loops are independent of each other and operate in parallel. However, this is not necessarily the case in an intact animal and under natural viewing conditions.

First of all, there are projections from the frontal eye fields to the superior colliculus (Lynch and Graybiel 1983) in addition to the projections to the paramedian pontine reticular formation (Leichnetz and Smith 1983). The functional nature of this connection is not quite clear. It may be that the frontal cortex controls the collicular pathway in such a way that reflex – like saccades (like express saccades) mediated by the first loop are inhibited. This notion is supported by the observation of Guitton et al. (1985), who noticed that patients with lesions of the frontal cortex cannot prevent reflexive glances to suddenly appearing stimuli and that these patients eventually execute express saccades.

The notion that the frontal eye fields have to do with the volitional aspect of the execution of a saccade is further substantiated by recording human cortical potentials (Kurtzberg and Vaughan 1982). Negative potentials preceding self-paced saccades, occur most prominently in the frontal and, to some extent also in the parietal regions but not in the occipital region.

5.2 Substantia Nigra

This control of the first loop by the second loop may be either direct, utilizing the direct pathway from the frontal eye fields to the colliculus, or indirect, or both. A good candidate for an indirect control is the substantia nigra pars reticularis.

First of all, the intermediate layers of the superior colliculus receive a direct projection from the substantia nigra (Hikosaka and Wurtz 1985a, where cells have high spontaneous activity which is suppressed prior to a visually guided saccade (Hikosaka and Wurtz 1981). The idea is that the superior colliculus is tonically inhibited by the activity coming from the substantia nigra and that only before a saccade this inhibition is reduced, allowing the colliculus to produce a presaccadic burst of spikes, which then initiates the saccade through the brain stem centers. Since it is known that gamma-aminobutyric acid (GABA) is the transmitter for that inhibition, it was possible to substantiate this idea by pharmacological experiments both in monkeys (Hikosaka and Wurtz 1985a; Hikosaka and Wurtz 1985b) and in cats (Joseph and Boussaoud 1985; Boussaoud and Joseph 1985). Injections of muscimol (a GABA agonist) or bicuculline (a GABA antagonist) into the superior colliculus clearly alter the oculomotor behavior of the animals. Unilateral injections of muscimol led to a suppression of contralateral saccades, particularly when the saccades had to be made to remembered target locations. In contrast, bicuculline injections facilitated the initiation of saccades. Of particular interest here is the fact that these animals produced apparently irrepressible saccades toward the movement fields of the collicular cells in the affected region.

It seems that the first loop is disinhibited in such a way that it cannot be prevented from producing saccades to visual stimuli. To demonstrate further that the GABA-related effects originate from the substantia nigra, muscimol was injected into the latter structure. The result was similar to that obtained after bicuculline injections in the superior colliculus (Hikosaka and Wurtz 1985b). The explanation is that GABA is also the transmitter of inhibitory synapses in the substantia nigra. Figure 5 shows schematically how the first loop can be modulated by the second loop through the nigral collicular pathway. The fact that both the frontal eye fields and the substantia nigra cells change their state of activity prior to saccades to a remembered target location further supports the idea that the two loops discussed so far can operate in parallel under certain circumstances but also in sequence, because the second loop can inhibit the first loop until the operation of the second loop (presumably decision-making even in cases where no target is visible)

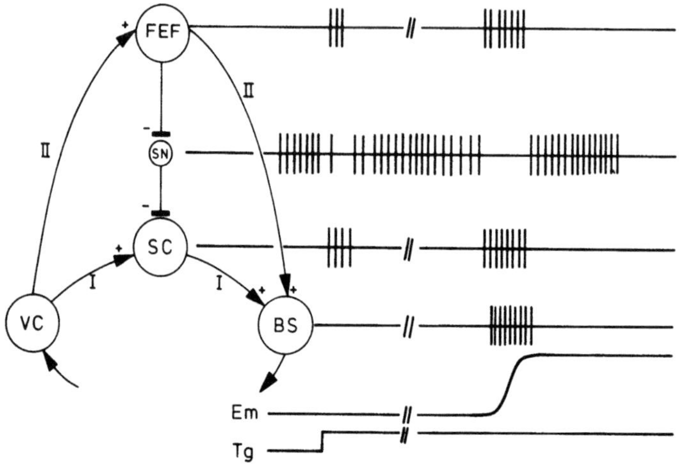

Fig. 5. A possible role of the substantia nigra (*SN*) in the control of visually guided saccades. The *SN* receives inhibitory impulses from the second loop, via the frontal eye fields (*FEF*), and inhibits the first loop, via the superior colliculus (*SC*). *V*, visual cortex; *BS*, brain stem. On the *right*, some corresponding impulse patterns of the different structures are shown

is completed. Only then can the first loop begin and complete its own operation (presumably computing saccade size and direction in space).

6 Third Loop – Fixation and Attention

When one thinks about the initiation of saccades, one also has to take into account the fact that in the absence of saccades the eyes can be fixated in such a way that the projection of a selected small target (stationary or slowly moving) stays at the fovea or very close to it. Little is known about how this is achieved, but it is clear that fixation is an active process, which must be interrupted before a saccade can occur.

Similarly, the initiation of voluntary goal – directed saccades should be closely related to shifts of visual attention because one usually wants to fixate targets of interest, which require a saccade to be positioned at the fovea.

The problem is: how are targets selected for attention and fixation, and how is this process of selection linked to the initiation of saccades?

In anticipation of the experimental results described below, Fig. 1 shows a third loop, which involves a process labeled A (attention). That a third loop exists at all, however, is concluded from reaction time studies in monkey and man. If subjects are forced attentively to fixate a small target, which remains visible when a new target appears in the periphery, saccadic reaction times are longest – in the order of 200 ms or more, as contrasted with the express saccades and the fast regular saccades (see Fig. 3). Further experiments in man have shown that the distribution of saccadic reaction times changes drastically according to the instructions "pay attention" or "do not pay attention", respectively (Mayfrank et al. 1986).

Although there is no generally accepted definition of what "attention" means, everybody knows from introspection that there exists an internal mechanism by which one can improve the perception of visual objects even if they are not fixated, i.e., if they are seen by peripheral vision (Singer et al. 1977; Bashinski and Bacharach 1980). That the attentional system is different from the localization system (C) was shown by Holtzman et al. (1981) on the basis of their observations of commissurotomy patients.

6.1 Saccadic Reaction Times

If, therefore, attention – regardless of its scientific definition – has anything to do with the initiation of saccades, one should be able to see changes in the reaction time when the direction of attention is manipulated by verbal instructions.

The results of a corresponding series of experiments by Mayfrank et al. (1986) are summarized in Fig. 6. Human subjects produced a bimodal distribution of their saccadic reaction times in a situation in which the fixation point disappeared some time before the peripheral target appeared (gap, upper left of Fig. 6). If the fixation point remained visible (overlap, upper right of Fig. 6), long latency saccades were elicited.

The data in B and C of Fig. 6 were obtained when the subjects were instructed to dissociate their direction of gaze from their direction of attention. In B, attention was directed to a small spot (physically identical to the fixation point used in A), while gaze direction was maintained at the center of the screen. If the attention target was turned off before the saccade target appeared (randomly on the right or left), a bimodal distribution was again obtained (Fig. 6B, left). If the attention target remained visible, saccades had long latencies (Fig. 6B, right).

This result shows that the attentional mechanism has somehow to become active before the saccade can be made, and this takes extra time (included in the reaction time) in the case of the overlap but not in the case of the gap. The question, however, remains open as to whether or not attention must already be directed to the new target location as a prerequisite for the saccade. One may, in fact, argue that the shift of

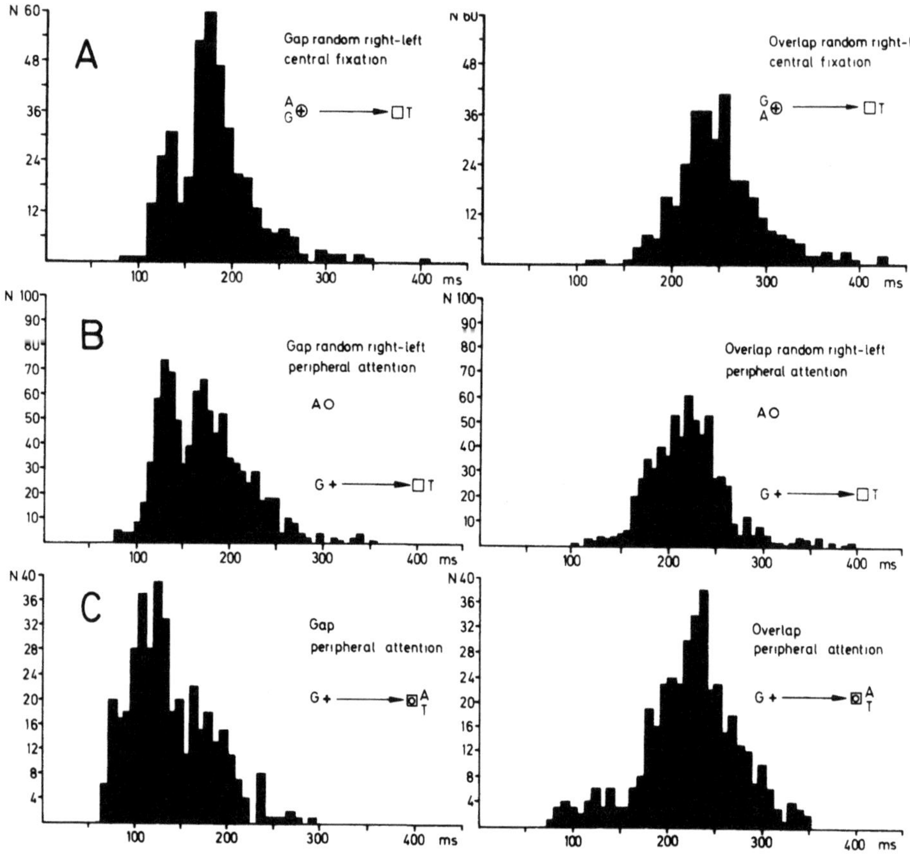

Fig. 6 A–C. Distribution of human saccadic reaction times in the gap (*left panel*) and overlap (*right panel*) experiment. **A** shows the usual result obtained when the direction of gaze (*G, cross*) coincides with the direction of visual attention (*A, open circle*). **B** shows about the same result, even though here the direction of gaze was dissociated from that of attention, which was directed to a position at the *upper vertical meridian*. **C** Here attention was directed to the target position (*T, open square*). Note: in the overlap paradigm (*lower right panel*) only very few express saccades occur

attention must precede the shift of gaze. The results shown in C of Fig. 6 give the answer: subjects were instructed to direct their attention to a target positioned at the location of the future saccade target such that shifts of attention were already completed in both situations. Yet short latency saccades were obtained only in the gap paradigm (Fig. 6C, left), whereas in the overlap paradigm saccades had long latencies (Fig. 6C, right). Mayfrank and his coworkers concluded from these observations that the mechanism of attention has two or even three aspects. The attentional system can be in two modes: one – the engaged mode – blocks the saccade system, and the other – the disengaged mode – enables but not necessarily produces saccadic eye movements.

Since attention can be directed successively to different locations in the visual field, one must assume that there exists a third mode, when attention moves. It should be pointed out here that this conclusion was reached independently by Posner and his group (Posner et al. 1984) on the basis of their observation of patients with parietal injuries. In fact, the terms "engage" and "disengage" have been adopted from Posner's papers (see below). The movement time of attention across the visual field was measured in man by Remington and Pierce (1984). They found that this time was constant.

6.2 Parietal Cortex

It was suggested long ago that the parietal lobe has to do with visual attention. Recent neurophysiological work has substantiated this idea. Lynch et al. (1977) were among the first to present data from single-unit recording in awake monkeys which suggested that, in fact, directed visual attention as a prerequisite for goal-directed (reach and eye) movements is one of the main factors governing the discharge patterns of neurons in the parietal cortex (area 7). Only later did it became clear that most of these cells also have access to visual information per se (Robinson et al. 1978; Mountcastle et al. 1981; Motter and Mountcastle 1981). The important observation is that the visual responses of these parietal cells can be modulated by the behavioral state of the animal with respect to the visual stimuli he sees in his field of vision. If, for example, a small light stimulus appeared, which the monkey had to foveate by a saccadic eye movement or to reach by a goal-directed arm movement, the passive visual on responses were enhanced (Bushnell et al. 1981); (Goldberg and Bruce 1985). If, on the other hand, the monkey was not fixating anything (eye pauses during intertrial intervals), the visual on responses were smaller or even abolished altogether (Mountcastle et al. 1981; Steinmetz et al. 1985).

The ultimate experiment, however, still needs to be described: this is to show that the cells' activity in the parietal cortex can be modulated in the absence of any change of light stimulation in the receptive field of the cells and in the absence of any movement. This experiment was carried out successfully by Fischer and Boch (1985) in the monkey's visual association cortex of the prelunate gyrus and will be considered below.

If one thinks of fixation as a situation in which the direction of gaze and the direction of visual attention coincide in space, one expects a neural structure which deals with attention also to deal with fixation. In fact, there are cells in the parietal cortex that discharge at high rates if the animal attentively fixates a small target. These fixation neurons described by Mountcastle and his group are silent during saccades. When the fixation point is positioned at different distances from the center and/or from the animal such that fixation of the target requires different positions of the eyes in the orbit, the discharge of the fixation neurons varies systematically (Sakata et al. 1980). It is not known, however, whether these changes are a consequence of changes in eye position or vice versa. In any case, one has to conclude that the cells in the parietal cortex receive extra retinal information that changes with the spatial location of the target which the monkey fixates and/or in which it is interested.

The contribution of the parietal cortex in the preparation of saccades is further elucidated by the observation that saccades elicited by electrical stimulation of this region are abolished if the monkey is actively fixating (Shibutani et al. 1984). This experiment not only shows that fixation is more than the absence of saccades but also suggests that the process of fixation inhibits the saccade system. This is equivalent to the notion that directed attention inhibits the saccade system, as has been concluded from the reaction time studies discussed above. Interestingly, saccades elicited electrically from the frontal eye fields are also reduced in amplitude, and thresholds are higher when the monkey attentively fixates, in contrast to a situation in which its eye is held still in the orbit (Goldberg et al. 1986).

If attention has these two or three modes and if the parietal cortex is involved in the sequences disengage – move – engage, one expects corresponding deficits after lesions in this cortical region.

Posner et al. (1984) conducted an extensive study of the effects of parietal injuries on covert orienting of attention. Their main finding was that the patients could not take advantage either of a peripheral or of a central cue to reduce their hand reaction time in response to a target in the visual field contralateral to the lesion. They concluded that the problem for such patients was to disengage attention from the central fixation area and they argued that the parietal cortex is needed for this disengagement. Damage to the parietal cortex in man not only leads to deficits in orienting visual attention but also results in a more general neglect syndrome which cannot be explained by sensory or motor deficits (Werth et al. 1986; Pierrot-Deseilligny et al. 1986).

The visual information that must reach the parietal cortex passes through the striate preoccipital cortical area for anatomical reasons as well as because of behavioral deficits in spatial vision (Mishkin and Ungerleider 1982). According to Fig. 1, process A sends information both to process D and to the oculomotor centers. If D is identified with the frontal eye fields and A with the parietal cortex, one expects anatomical connections between them. In fact, Schwartz and Goldman-Rakic (1984) have shown that area 8A receives a clear ipsilateral projection from the intraparietal sulcus.

6.3 Extrastriate Cortex

There is no reason to believe that the parietal cortex is the only structure involved in directing visual attention. Rather, one has to assume that any structure which projects to parietal areas may be the source of visual attention and that any structure which receives projection from the parietal cortex might be influenced by visual attention. In fact, Fig. 1 suggests that processes D and C as well as M are under the control of the attention system, an aspect that will be considered later.

On the other hand, any structure that receives visual information (process V in Fig. 1) and connects somehow to the oculomotor centers is a candidate for visual attention in the context of visually guided saccades. The extrastriate cortex is such a structure, particularly the prelunate gyrus. Area MT is not considered in this context because it seems to be involved in the interaction of motion and slow pursuit eye movements rather than saccadic eye movements (Newsome et al. 1985). Nevertheless, since saccades must be suppressed during slow pursuit as well as during fication, the saccade system and the slow pursuit system must also interact. It is not surprising, therefore, to see connections between area MT and saccade-related structures (Ungerleider et al. 1984; Maioli et al. 1983a; Maioli et al. 1983b).

In an attempt to demonstrate the effects of attention and fixation on the visual responses of cells in the prelunate cortex of the rhesus monkey, Fischer and Boch (1985) trained monkeys to maintain stationary fixation of a small spot even in the presence of a peripheral target. When the fixation point was turned off, the animals were free to move or not to move their eyes to the peripheral stimulus; all they had to do was to detect correctly a small dimming of the luminance of the stimulus. The result is summarized in Fig. 7. Regardless of whether or not the monkey made a saccade, many prelunate cells were activated after the offset of the fixation point even though there was no change of visual stimulation in the receptive fields. This reactivation was interpreted as a sign of the animal's having directed its attention to the peripheral target in order to detect its dimming.

Unfortunately, this paradigm has not yet been used to test neurons in the parietal or frontal cortex or in the superior colliculus, and it is therefore still not known whether or not the absence of a fixation point changes the visual properties of the cells in these structures when

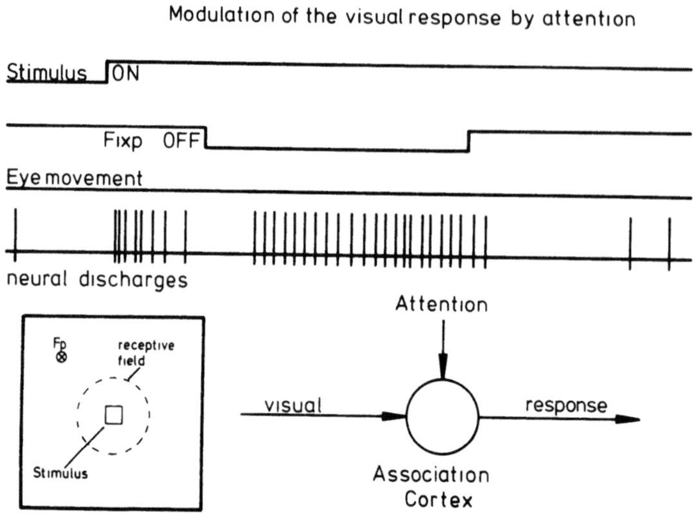

Fig. 7. Discharge pattern of a neuron in the prelunate gyrus of the rhesus monkey. The neuron gives a moderate on response when a stimulus appears in its receptive field (*lower left inset*). When the fixation point is switched off some time later, the neuron begins to discharge again at a higher rate even though the monkey suppresses the saccade to the stimulus

the monkey makes no eye movements. These questions, however, seem to be under study now in several groups (Moran and Desimone 1985; Steinmetz et al. 1985). It is known what changes of neuronal activity occur in all of these structures before visually guided saccades to suddenly appearing targets take place.

6.4 Presaccadic Enhancement

According to Fig. 1, the attentional system (irrespective of where and how it is organized) feeds into the second loop, which then feeds into the first loop. If this is indeed the case, one expects any structure involved in the three loops to change its activity if the visual stimulus, used to activate a cell, is a target of a saccade. This presaccadic modulation of the visual response to the onset of a light stimulus was termed the "enhancement effect" by Wurtz and Goldberg (1972) and Wurtz and Mohler (1976b) when they first reported the effect in the superior colliculus. Later, the enhancement effect was demonstrated in the frontal eye fields (Goldberg and Bushnell 1981), the parietal cortex (Robinson et al. 1978), and the prelunate gyrus (Fischer and Boch 1981a). The scheme shown in Fig. 1 suggests that the enhancement effects in all these structures are indeed due to and originate from a single mechanism, namely the mechanism of visual attention. The enhancement effect has also been demonstrated in the pulvinar (Petersen et al. 1985), as well as in a small percentage of cells in the striate cortex (Wurtz and Mohler 1976b), with the important difference that the effect was not spatially selective but rather occurred before saccades in any direction.

7 Links Between Loops

Figure 1 seems to suggest that the three loops may operate independently and in parallel. But the effect of the linkage between the loops, as outlined in the diagram, may be that the three processes must take place one after the other in order to produce a correct saccade. For example, if processes A and D are completed but C is not (because the saccade target has not yet occurred), a saccade may be initiated because the loops through A and D may drive the brain stem centers. However, this saccade would be wrong in amplitude and/or direction. This happens, for example, in the gap paradigm when monkeys or human subjects anticipate the occurrence of the target. They produce a premature saccade, which misses the target location and often seems to have a lower velocity as well.

Similarly, saccades can still be made even if process C cannot take place because the corresponding structure has been destroyed experimentally. But these saccades would fail to reach the target accurately. The anatomical evidence for the existence of the loops and their linkage is clear. Figure 8 outlines the most prominent anatomical connections. If one accepts the identifications proposed above, both the parietal cortex and the extrastriate cortex project to the frontal eye fields (Petrides and Pandya 1984; Schwartz and Goldman-Rakic 1984), and the

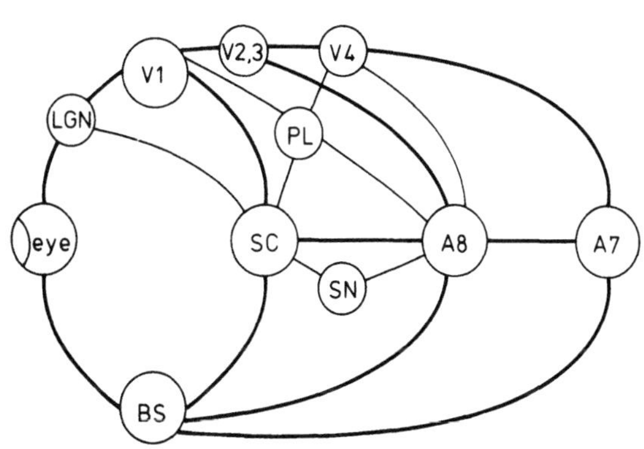

Fig. 8. Anatomical connections between different structures involved in the preparation of visually guided saccades. The *heavy lines* reflect the three-loop concept from Fig. 1

frontal eye fields project to the superior colliculus (Leichnetz et al. 1981) directly and indirectly through the substantia nigra.

The projection from the parietal cortex to the brain stem does not necessarily imply that activation of this pathway directly leads to saccadic eye movements. On the contrary, the occurrence of saccades through electrical stimulation of the extrastriate and parietal cortex critically depends on an intact superior colliculus (Keating et al. 1983). The pathway from A7 to the colliculus may functionally include the frontal eye fields (Barbas and Mesulam 1981).

Figure 8, even though simplified and incomplete, looks rather confusing. In fact, if one were to continue to fill in all the known anatomical connections, the figure would be almost useless. However, if one groups together the different visual cortical and subcortical structures, the three-loop concept can easily be visualized. What remains to be investigated is how these connections exactly work in terms of excitation and inhibition.

8 Modifications by Practice and Learning

The saccadic system is used from early childhood and certainly undergoes several modifications during ontogenetic development (Bronson 1981). Here we consider the possibility that the preparatory processes and their sequential or parallel operation can be modulated in adults. Again, any changes either in the individual stages or in their cooperation should be reflected in the corresponding reaction time.

As well as observing that the presaccadic activation in extrastriate cortical cells was largely reduced after 10–20 days of daily practice, Fischer and Boch (1982) noticed that the saccadic reaction times of the monkeys were reduced from about 400 ms at the beginning to about 220 ms at the end of a training period (Boch and Fischer 1983). When they used the gap paradigm, monkeys produced only a few express saccades in the beginning but increased the amount to almost 100% by a week or two of daily practice (Fischer et al. 1984). Similarly, it was possible to increase the number of express saccades from almost zero to 50% if the overlap paradigm was used in the monkey (Fischer et al. 1984). More specifically, the percentage of express saccades was increased only for target positions that had been used during the training periods but not for others. In other words, the effect of daily practice turned out to be spatially selective in the monkey.

The notion of three loops offers two possible explanations:

1. One or other of the participating processes decreases its (their) time for completion. This would result in a steady and continuous shift of the corresponding peaks of the distribution of the saccadic reaction times.
2. The effect of practice is that an increasing number of saccades are executed through the use of the faster loops. For example, using the overlap paradigm in naive subjects results in long reaction times. By practice they learn to produce an increasing number of express saccades and of fast regular saccades. The overall mean value shifts toward shorter values, but what in fact changes are the relative heights of the peaks in the distributions.

Figure 9 illustrates the effects of daily practice by a single human subject, who made 200 right-directed saccades per day in the overlap paradigm. In the beginning the distribution seems to have three peaks: a small one between 100 and 150 ms, a second below 200 ms, and a large broad peak above 200 ms. Saccades to the left are hardly different from those to the right (Fig. 9, upper right and left). As practice progresses, the first and second peaks increase, whereas the third decreases. Finally, the third peak, which was so pronounced in the beginning, has almost disappeared altogether. Saccades to the left, the untrained side, were about the same as in the beginning. The subject now produced a completely different distribution of reaction times in the two directions (Fig. 9, lower right and left). The data of Fig. 9 favor the second explanation, i.e., what changes is the relative contribution of the three loops to the distribution of the reaction times. Since it is the third peak that is so drastically reduced and since this peak includes the process of disengagement of attention (see Sect. 6), it may be concluded that practice enables the subject to disengage directed attention from the fixation point without an external stimulus. If this is the case, the initial distribution (untrained situation) should be restored immediately if one asks the subject attentively to fixate the fixation point. The corresponding experiment proves this prediction.

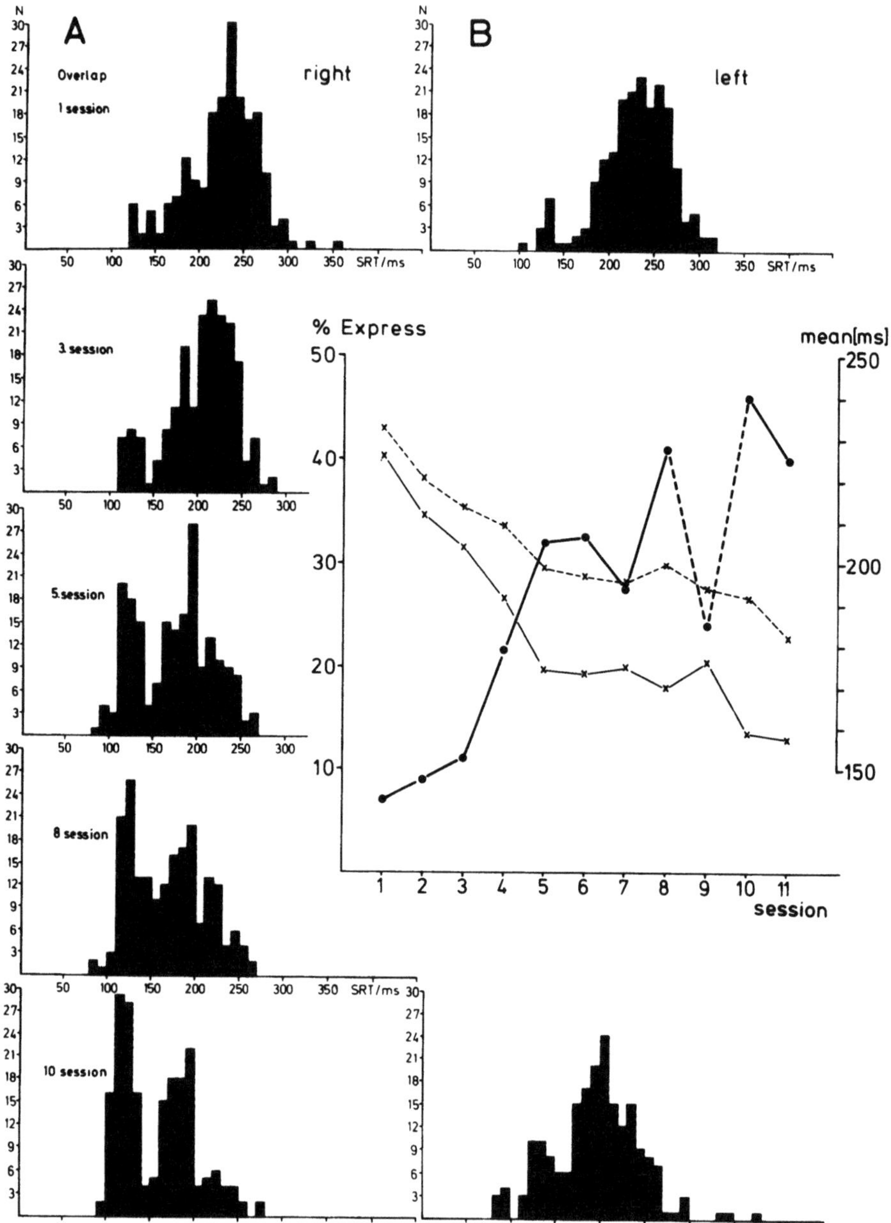

Fig. 9A, B. Effects of daily practice on the reaction times in the overlap paradigm. **A** shows the distributions as practice increased (from *top* to *bottom*). During the training sessions only saccades to the right were made. The control for saccades to the *left* is shown in **B** at the beginning (*top*) and at the end (*bottom*) of the training period of 10 days. The *inset* shows the increase in the percentage of express saccades (*heavy line*) and the decreases in the mean reaction time. *Broken line,* mean of all saccades except express; *thin line,* mean of all saccades

9 Conclusion

From our present knowledge it follows that a complete preparation of a single saccade includes at least three processes which altogether take about 200 ms or even a little more. Under natural viewing conditions one makes about two to three saccades in a second, so that the time between the end of one saccade and the beginning of the next – the so-called fixation period – is just in the order of the reaction time. During this time visual information must be distributed to different parts of the brain. If one distinguishes mainly two systems, one for identification of visual objects by foveal inspection and one for localization of objects in visual space by peripheral vision, as discussed by Mishkin and Ungerleider (1983) and Sagi and Julesz (1985), it seems reasonable to feed these two systems sequentially during the time of no saccades. Figure 10 outlines this idea. It utilizes the notion that visual attention has two stages – engagement and disengagement – and that disengagement is necessary for a shift of the direction of attention as well as for a shift of the direction of gaze.

The transition from the engaged to the disengaged state takes place some time in the middle of a saccade-free period. During the engaged period visual information is made available for the identification system (probably including the inferotemporal cortex), and during the disengaged period the localization system receives visual information, which can than be used to guide a movement, e.g., of the eye and/or the hand, or the direction of attention without any other movement, as indicated by the horizontal broken line in the eye movement trace of Fig. 10.

An important point in this notion is that the identification and localization systems can work in parallel, the only sequential operation being the loading of visual information into the two systems. This notion also explains why effects of attention can be seen in both systems,

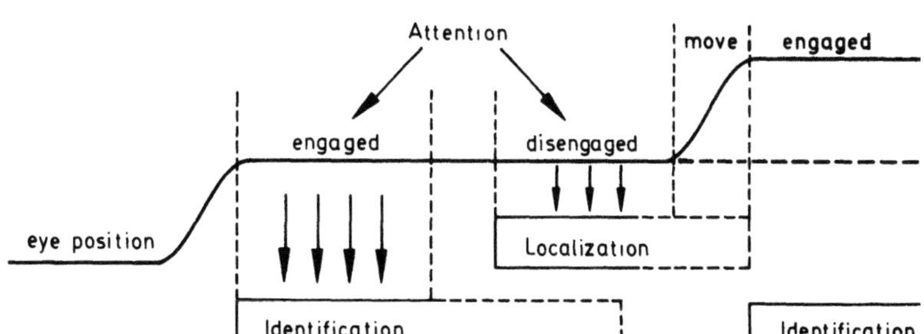

Fig. 10. Sequence of two saccades and the related changes in the state of attention

particularly in the identification system, presumably involving the inferotemporal cortex (Nuwer and Pribram 1979; Mikami et al. 1982), and it explains furthermore why the effects of attention may be different – even opposite in sign – in the parietal as compared to the inferotemporal cortex (Richmond et al. 1983).

Fig. 11 shows how the systems of identification (I), localization (C), attention (A), and decision-making (D) may be linked together. It will be noted that the part marked with heavy lines is identical with Fig. 1. Here two visual subsystems are considered. VF is intended to represent the visual features of objects, which by allowance of the attentional system (engage) are fed into the identification system, whereas VP represents a map of visual positions, which by allowance of the attentional system (disengage) are fed into the localization system. Also added are boxes for central representation (cr) and memory (mem), which mediate the influence of the instructions and elaborate the report of a subject according to the results of I and C. The diagram is a highly simplified hypothetical concept of how different aspects of vision and visually guided movements may be organized and how the attentional system may play a central role in the coordination.

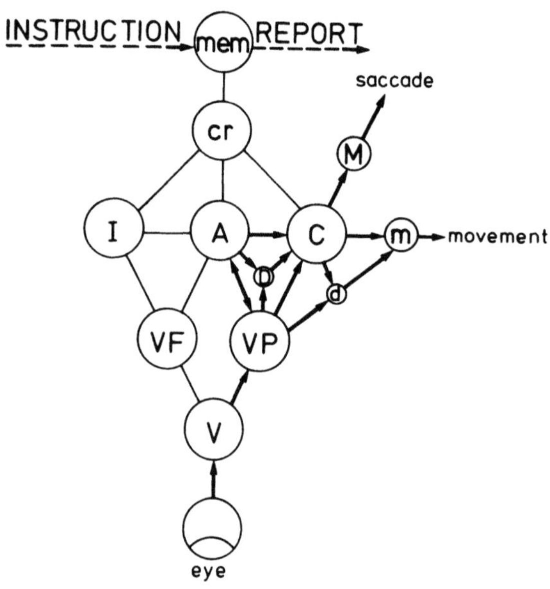

Fig. 11. Possible coordination of subsystems for identification (*I*) and localization (*C*) by attention (*A*). Note that the *heavy lines* represent again the three-loop concept for the preparation of a saccade. The localization system can be used to initiate other visually guided movements (*m*), which of course need their own decision mechanism (*d*). The scheme includes two visual structures, one for features representation (*VF*) and one for a position map (*VP*). Also included are a central representation (*cr*), which combines identified features and selected positions and eventually feeds into a memory (*mem*)

References

Barbas H, Mesulam MM (1981) Organization of afferent input to subdivisions of area 8 in the rhesus monkey. J Comp Neurol 200:407–431
Bashinski HS, Bacharach VR (1980) Enhancement of perceptual sensitivity as the result of selectively attending to spatial locations. Percept Psychophys 28: 241–248
Becker W, Juergens R (1979) An analysis of the saccadic system by means of double step stimuli. Vis Res 19:967–983
Bizzi E (1968) Discharge of frontal eye field neurons during saccadic and following eye movements in unanesthetized monkeys. Exp Brain Res 6:69–80
Boch R, Fischer B (1983) Saccadic reaction times and activation of prelunate cortex: parallel observations in trained rhesus monkeys. Exp Brain Res 50:201–210
Boch R, Fischer B (1986) Further observations on the occurrence of express saccades in monkey. Exp Brain Res 63:487–494
Boch R, Fischer B, Ramsperger E (1984) Express-saccades of the monkey: reaction times versus intensity, size, duration, and eccentricity of their targets. Exp Brain Res 55:223–231
Boch R (1986) Unpublished observation and personal communication
Bolz J, Rosner G, Waessle H (1982) Response latency of brisk-sustained (X) and brisk-transient (Y) cells in the cat retina. J Physiol 328:171–190
Boussaoud D, Joseph JP (1985) Role of the cat substantia nigra pars reticulata in eye and head movements. II. Effects of local pharmacological injections. Exp Brain Res 57:297–304
Bronson GW (1981) The scanning patterns of human infants implication for visual learning. In: Lipsitt LP (ed) monographs on infancy. Ablex, Norwood NJ
Bruce CJ, Goldberg ME (1984) Physiology of the frontal eye fields. Trends Neurosci 7:436–441
Bruce CJ, Goldberg ME (1985) Primate frontal eye fields. I. Single neurons discharging before saccades. J Neurophysiol 53:603–635
Bruce CJ, Goldberg ME, Bushnell MC, Stanton GB (1985) Primate frontal eye fields. II. Physiological and anatomical correlates of electrically evoked eye movements. J Neurophysiol 54:714–734
Bushnell MC, Goldberg ME, Robinson DL (1981) Behavioral enhancement of visual responses in monkey cerebral cortex. I. Modulation in posterior parietal cortex related to selective visual attention. J Neurophysiol 46:755–772
Campos-Ortega JA, Hayhow WR (1972) On the organization of the visual cortical projection to the pulvinar in macaca mulatta. Brain Beh Evol 6:394–423
Collin NG, Cowey A (1980) The effect of frontal eyefields and superior colliculi on visual stability and movement discrimination in rhesus monkey. Exp Brain Res 40:251–260
Coren S, Hoenig P (1972) Effect of non-target stimuli upon length of voluntary saccades. Percept Mot Skills 34:499–508
Desimone R, Moran J (1985) Mechanisms for selective attention in area V4 and inferior temporal cortex of the macaque. Neurosci Abstr 364. 8:1245
Deubel H, Wolf H, Hauske G (1982) Corrective saccades: effect of shifting the saccade goal. Vis Res 22:353–364
Findlay JM (1982) Global visual processing for saccadic eye movements. Vis Res 22:1033–1045
Fischer B, Boch R (1981a) Enhanced activation of neurons in prelunate cortex before visually guided saccades of trained rhesus monkeys. Exp Brain Res 44: 129–137
Fischer B, Boch R (1981b) Selection of visual targets activates prelunate cortical cells in trained rhesus monkeys. Exp Brain Res 41:431–433

Fischer B, Boch R (1982) Modifications of presaccadic activation of neurons in the extratriate cortex during prolonged training of rhesus monkeys in a visuo-oculomotor task. Neurosci Lett 30:127–131
Fischer B, Boch R (1983) Saccadic eye movements after extremely short reaction times in the monkey. Brain Res 260:21–26
Fischer B, Boch R (1985) Peripheral attention versus central fixation: modulation of the visual activity of prelunate cortical cells of the rhesus monkey. Brain Res 345:111–124
Fischer B, Boch R, Bach M (1981) Stimulus versus eye movements: comparison of neural activity in the striate and prelunate visual cortex (A17 and A19) of trained rhesus monkey. Exp Brain Res 43:69–77
Fischer B, Boch R, Ramsperger E (1984) Express-saccades of the monkey: effects of daily training on probability of occurrence and reaction time. Exp Brain Res 55:232–242
Fischer B, Ramsperger E (1984) Human express-saccades: extremely short reaction times of goal directed eye movements. Exp Brain Res 57:191–195
Fischer B, Ramsperger E (1986) Human express-saccades: effects of daily practice and randomization. Exp Brain Res 64:569–578
Frost D, Poeppel E (1976) Different programming modes of human saccadic eye movements as a function of stimulus eccentricity: indication of a functional subdivision of the visual field. Biol Cybern 23:39–48
Gentilucci M, Scandolare C, Pigarev IN, Rizzolatti G (1983) Visual responses in the postarcuate cortex (area 6) of the monkey that are independent of eye positon. Exp Brain Res 50:464–468
Goldberg ME, Bruce CJ (1981) Frontal eye fields in the monkey: eye movements remap the effective coordinates of visual stimuli. Neurosci Abstr 44. 4:131
Goldberg ME, Bruce CJ (1985) Cerebral cortical activity associated with the orientation of visual attention in the rhesus monkey. Vis Res 25:471–481
Goldberg ME, Bushnell MC (1981) Behavioral enhancement of visual responses in monkey cerebral cortex. II. Modulation in frontal eye fields specifically related to saccades. J Neurophysiol 46:773–787
Goldberg ME, Bushnell MC, Bruce CJ (1986) The effect of attentive fixation on eye movements evoked by electrical stimulation of the frontal eye fields. Exp Brain Res 61:579–584
Guitton D, Buchtel HA, Douglas RM (1985) Frontal lobe lesions in man cause difficulties in suppressing reflexive glances and in generating goal-directed saccades. Exp Brain Res 58:455–472
Guthrie BL, Porter JD, Sparks DL (1983) Corollary discharge provides accurate eye position information to the oculomotor systems. Science 221:1193–1195
Hering E (1879) Der Raumsinn und die Bewegungen der Augen. In: Hermann L (ed) Handbuch der Physiologie 3. Vogel, Leipzig, pp 343–601
Hikosaka O, Wurtz RH (1981) The role of substantia nigra in the initiation of saccadic eye movements. In: Fuchs AF, Becker W (eds) Elsevier North Holland, Amsterdam
Hikosaka O, Wurtz RH (1982) Visual fixation suppresses visual response of cells in monkey substantia nigra pars reticulata (ARVO 1982). Invest Ophthalmol Vis Sci 22:238
Hikosaka O, Wurtz RH (1983) Effects on eye movements of a GABA agonist and antogonist injected into monkey superior colliculus. Brain Res 272:368–372
Hikosaka O, Wurtz RH (1985a) Modification of saccadic eye movements by GABA-related substances. I. Effect of muscimol and bicuculline in monkey superior colliculus. J Neurophysiol 53:266–291
Hikosaka O, Wurtz RH (1985b) Modification of saccadic eye movements by GABA-related substances. II. Effects of muscimol in monkey substantia nigra pars reticulata. J Neurophysiol 53:292–308

Hochstein S, Maunsell JHR (1985) Dimensional attention effects in the responses of V4 neurons of the macaque monkey. Neurosci Abstr 364. 6:1244
Holtzmann JD, Sidtis JJ, Volpe BT, Wilson DH, Gazzaniga MS (1981) Dissociation of spatial information for stimulus localization and the control of attention. Brain 104:861–872
Jay MF, Sparks DL (1984) Auditory receptive fields in primate superior colliculus shift with changes in eye position. Nature 309:345–347
Jonides J, Irwin DE, Yantis S (1982) Integrating visual information from successive fixations. Science 215:192–194
Joseph JP, Boussaoud D (1985) Role of the cat substantia nigra pars reticulata in eye and head movements. I. Neural activity. Exp Brain Res 57:286–296
Jung R (1972) Introduction. – Conclusions: how do we see with moving eyes? Bibl Ophthal 82:1–6; 377–395
Keating EG, Gooley SG, Pratt SE, Kelsey JE (1983) Removing the superior colliculus silences eye movements normally evoked from stimulation of the parietal and occipital eye fields. Brain Res 269:145–148
Kurtzberg D, Vaughan Jr HG (1982) Topographic analysis of human cortical potentials preceding self-initiated and visually triggered saccades. Brain Res 243:1–9
Leichnetz GR, Spencer RF, Hardy SGP, Astruc J (1981) The prefrontal cortico tectal projection in the monkey: an anterograde and retrograde horseradish peroxidase study. Neurosci 6:1023–1041
Leichnetz GR, Smith DJ (1983) Prefrontal, frontal eye field, and area 6 projections to the paramedian pontine reticular formation (PPRF) in the monkey. Neurosci Abstr 220. 4:749
Lynch JC, Graybiel AM (1983) Comparison of afferents traced to the superior colliculus from the frontal eye fields and from two sub-regions of area 7 of the rhesus monkey. Neurosci Abstr 220. 5:750
Lynch JC, Mountcastle VB, Talbot WH, Yin TCT (1977) Parietal lobe mechanisms for directed visual attention. J Neurophysiol 40:362–389
MacKay DM (1973) Visual stability and voluntary eye movements. In: Jung R (ed) Central processing of visual information. Springer, Berlin Heidelberg New York, pp 307–331 (Handbook of sensory physiology, vol 7/3)
Maioli MG, Squatrito S, Galletti C, Battaglini PP, Sanseverino ER (1983) Cortico-cortical connections from the visual region of the superior temporal sulcus to frontal eye field in the macaque. Brain Res 265:294–299
Mayfrank L, Mobashery M, Kimmig H, Fischer B (1986) The role of fixation and visual attention on the occurrence of express saccades in man. Eur J Psychiatr Neurol Sci 235:269–275
Mays LE, Sparks DL (1980a) Dissociation of visual and saccade-related responses in superior colliculus neurons. J Neurophysiol 43:207–232
Mays LE, Sparks DL (1980b) Saccades are spatially, not retinocentrically, coded. Science 208:1163–1165
Mikami A, Ito S-I, Kubota K (1982) Modifications of neuron activities of the dorso-lateral prefrontal cortex during extrafoveal attention. Behav Brain Res 5:219
Mishkin M, Ungerleider LG (1982) Contribution of striate inputs to the visuospatial functions of parietoprecoccipital cortex in monkeys. Behav Brain Res 6:57–77
Mishkin M, Ungerleider LG, Macko KA (1983) Object vision and spatial vision: two cortical pathways. Trends Neurosci 6:414–417
Moran J, Desimone R (1985) Selective attention gates visual processing in area V4 and the inferior temporal cortex of the macaque. Neurosci Abstr 364. 7:1245
Motter BC, Mountcastle VB (1981) The functional properties of the lightsensitive neurons of the posterior parietal cortex studied in waking monkeys – foveal sparing and opponent vector organization. J Neurosci 1:3
Mountcastle VB, Andersen RA, Motter BC (1981) The influence of attentive fixation upon the excitability of the light-sensitive neurons of the posterior parietal cortex. J Neurosci 1:1218–1235

Newsome WT, Wurtz RH, Duersteler MR, Mikami A (1985) Deficits in visual motion processing following ibotenic acid lesions of the middle temporal visual area of the macaque monkey. J Neurosci 5:825–840
Newsome WT, Wurtz RH, Duersteler MR, Mikami A (1985) Punctate chemical lesions of striate cortex in the macaque monkey: effect on visually guided saccades. Exp Brain Res 58:393–399
Nuwer MR, Pribram KH (1979) Role of the inferotemporal cortex in visual selective attention. Electroenceph Clin Neurophysiol 46:389–400
Petersen SE, Robinson DL, Keys W (1985) Pulvinar nuclei of the behaving rhesus monkey: visual responses and their modulation. J Neurophysiol 54:867–886
Petrides M, Pandya DN (1984) Projections to the frontal cortex from the posterior parietal region in the rhesus monkey. J Comp Neurol 228:105–116
Pierrot-Deseilligny C, Gray F, Brunet P (1986) Infarcts of both inferior parietal lobules with impairment of visually guided eye movements, peripheral visual inattention and optic ataxia. Brain 109:81–97
Posner MI, Walker JA, Friedrich FJ, Rafal RD (1984) Effects of parietal injury on covert orienting of attention. J Neurosci 4:1863–1874
Remington R, Pierce L (1984) Moving attention: evidence for time-invariant shifts of visual selective attention. Percept Psychophys 35:393–399
Reulen JPH (1984) Latency of visually evoked saccadic eye movements. I. Saccadic latency and the facilitation model. Biol Cybern 50:251–262
Reulen RPH (1984) Latency of visually evoked saccadic eye movements. II. Temporal properties of the facilitation mechanism. Biol Cybern 50:263
Richmond BJ, Wurtz RH (1980) Vision during saccadic eye movements. II. A corollary discharge to monkey superior colliculus. J Neurophysiol 43:1156–1167
Richmond BJ, Wurtz RH, Sato (1983) Visual responses of inferior temporal neurons in awake rhesus monkey. J Neurophysiol 50:1415–1432
Rizzolatti G, Scandolara C, Matelli M, Gentilucci M (1981) Afferent properties of periarcuate neurons in macaque monkeys. II. Visual responses. Behav Brain Res 2:147–163
Robinson DA (1972) Eye movements evoked by collicular stimulation in the alert monkey. Vis Res 12:1795–1808
Robinson DL, Baizer JS, Dow BM (1980) Behavioral enhancement of visual responses of prestriate neurons of the rhesus monkey. Invest Ophthalmol Vis Sci 19: 1120–1123
Robinson DL, Goldberg ME, Stanton GB (1978) Parietal association cortex in the primate: sensory mechanisms and behavioral modulations. J Neurophysiol 41:910–932
Robinson DL, Petersen SE (1985) Responses of pulvinar neurons to real and self-induced stimulus movement. Brain Res 338:392–394
Robinson DL, Wurtz RH (1976) Use of an extraretinal signal by monkey superior colliculus neurons to distinguish real from self-induced stimulus movements. J Neurophysiol 39:852–870
Sagi D, Julesz B (1985) "Where" and "what" in vision. Science 228:1217–1219
Sakata H, Shibutani H, Kawano K (1980) Spatial properties of visual fixation neurons in posterior parietal association cortex of the monkey. J Neurophysiol 43: 1654–1672
Sandell JH, Schiller PH, Maunsell JHR (1984) The effect of superior colliculus and frontal eye field lesions on saccadic latency in the monkey. Perception 13:A6
Saslow MG (1967) Effects of components of displacement-step stimuli upon latency of saccadic eye movements. J Opt Soc Am 57:1024–1029
Schiller PH (1977) The effect of superior colliculus ablation on saccades elicited by cortical stimulation. Brain Res 122:154–156
Schiller PH, True SD, Conway JL (1979) Paired stimulation of the frontal eye fields and the superior colliculus of the rhesus monkey. Brain Res 179:162–164

Schiller PH, Sandell JH (1983) Interactions between visually and electrically elicited saccades before and after superior colliculus and frontal eye field ablations in the rhesus monkey. Exp Brain Res 49:381–392

Schiller PH, Stryker M (1972) Single-unit recording and stimulation in superior colliculus of the alert rhesus monkey. J Neurophysiol 35:179–196

Schiller PH, True SD, Conway JL (1979) Paired stimulation of the frontal eye fields and the superior colliculus of the rhesus monkey. Brain Res 179:162–164

Schlag J, Schlag-Rey M (1984) Visuomotor functions of central thalamus in monkey. II. Unit activity related to visual events, targeting, and fixation. J Neurophysiol 51:1175–1195

Schlag J, Schlag-Rey M (1985) Eye fixation units in the supplementary eye field of monkey. Neurosci Abstr 25. 23:82

Schwartz ML, Goldman-Rakic PS (1984) Callosal and intrahemispheric connectivity of the prefrontal association cortex in rhesus monkey: relation between intraparietal and principal sulcal cortex. J Comp Neurol 226:403420

Shibutani H, Sakata H, Hyvaerinen J (1984) Saccade and blinking evoked by microstimulation of the posterior parietal association cortex of the monkey. Exp Brain Res 55:1–8

Singer W, Zihl J, Poeppel E (1977) Subcortical control of visual thresholds in humans: evidence of modality specific and retinotopically organized mechanisms of selective attention. Exp Brain Res 29:173–190

Sparks DL, Mays LE (1983) Spatial localization of saccade targets. I. Compensation for stimulation-induced perturbations in eye position. J Neurophysiol 49:45–63

Sparks DL, Porter JD (1983) Spatial localization of saccade targets. II. Activity of superior colliculus neurons preceding compensatory saccades. J Neurophysiol 49:64–74

Steinmetz MA, Motter BC, Mountcastle VB (1985) Attentive fixation influences differentially the responses of visual neurons of prestriate and parietal areas of the cerebral cortex. Neurosci Abstr 297/9:1012

Ungerleider LG, Mishkin M (1982) Two cortical visual systems. In: Ingle DJ, Goodale MA, and Nansfield RJW (eds) Analyses of visual behavior. MIT Press, Cambridge, pp 549–586

Wagman JH, Krieger HP, Bender MB (1958) Eye movements elicited by surface and depth stimulation of the occipital lobe of macaca mulatta. J Comp Neurol 109:169–193

Werth R, von Cramon D, Zihl J (1986) Neglect: Phänomene halbseitiger Vernachlässigung nach Hirnschädigung. Fortschr Neurol Psychiat 54:21–32

Wurtz RH (1969) Comparison of effects of eye movements and stimulus movements on striate cortex neurons of the monkey. J Neurophysiol 32:987–994

Wurtz RH, Albano JE (1980) Visual-motor function of the primate superior colliculus. Ann Rev Neurosci 3:189–226

Wurtz RH, Goldberg ME (1972) The primate superior colliculus and the shift of visual attention. Invest Ophthalmol 11:441–450

Wurtz RH, Goldberg ME, Robinson DL (1980) Behavioral modulation of visual responses in the monkey: stimulus selection for attention and movement. Progr Psychobiol Physiol Psychol 9:43–83

Wurtz RH, Mohler CW (1976a) Organization of monkey superior colliculus: enhanced visual response of superficial layer cells. J Neurophysiol 39:745–762

Wurtz RH, Mohler CW (1976b) Enhancement of visual responses in monkey striate cortex and frontal eye fields. J Neurophysiol 39:766–772

Rev. Physiol. Biochem. Pharmacol, Vol. 106

Gaze Stabilization in the Primate

The Interaction of the Vestibulo-Ocular Reflex, Optokinetic Nystagmus, and Smooth Pursuit

WALTER WAESPE and VOLKER HENN

Contents

Neurology Department, University of Zürich, 8091 Zürich, Switzerland
Supported by Swiss National Foundation for Scientific Research 3.593-0.84 and 3.510-0.86

1 Introduction

For precise vision the eyes have to be stabilized relative to a visual target or space during active or passive head movements. Gaze stabilization is accomplished by the interaction of inputs from several sensory systems, the vestibular, visual, and proprioceptive systems being the most important ones. This information is combined and transformed centrally into motor command signals to induce or suppress eye movements for gaze stabilization. One of several aims is to analyze and to understand how and where the transformation of sensory signals into motor signals takes place.

Under natural conditions the task of gaze stabilization is very complex, because the head is moved actively and usually against the body and because movements are in all three dimensions of space (Mach 1875). During the recording of activity from single neurons in the laboratory, a general reduction in the complexity of the movements was achieved by allowing only passive head movements, by moving the head and the body as a whole, and by restricting movements to one (usually the horizontal) plane.

The vestibulo-ocular reflex and optokinetic nystagmus have been analyzed in many species covering a wide range of phylogeny. The phenomenon of gaze stabilization and its basic mechanisms seem to be remarkably similar in different species (Sandeman 1983). The basic design of the vestibular system can be followed through phylogeny with modifications of connections to the oculomotor system in lateral-eyed animals like the rabbit and frontal-eyed animals like the monkey (Simpson and Graf 1985). In contrast, vision has evolved in a more complex way with the addition of the fovea. Vision in primates is fovea-dominated. Primates can voluntarily direct their visual attention towards single objects in space. Therefore gaze stabilization relies on information from the fovea and the extrafoveal retina. Problems of visual attention and target selection for eye movements have been summarized by Jung (1978) and Wurtz and coworkers (Wurtz et al. 1982). The neuronal mechanisms involved remain largely unknown. Neurons in the parietal cortex or frontal eye fields are modulated for instance depending on whether the animal is going to perform a saccade on the visual target or not (Lynch et al. 1977; Bushnell et al. 1981).

The decisive but not exclusive requirement for gaze stabilization is the detection of image shift on the retina. Many lower animals like the frog have retinal ganglion cells which are already specialized for detecting motion. In higher animals the tendency prevails that specialized information like motion is extracted in central structures. The basic information flow is the same: position changes of retinal images induce a velocity signal. The extraction of a velocity signal is the precondition that motion information from different sensory systems can interact. On the output side, motoneurons of the oculomotor system also need information about eye position. Mathematically, to obtain position information, velocity signals have to be integrated (Skavenski and Robinson 1973). This process will not be considered in this context.

In our review we will limit ourselves to brainstem mechanisms. The problem of cortical information processing and of voluntarily directing attention will not be discussed (cf, Fischer, this vol.). Adaptation phenomena and plastic changes have recently been reviewed (Berthoz and Melvill-Jones 1985).

We will first describe the phenomenology of the vestibulo-ocular reflex, optokinetic nystagmus, smooth pursuit, and their immediate interactions. We will restrict this descripton to eye movements in the horizontal plane. In the following chapters results of single neuron recordings, mainly from rhesus monkeys, will be described. Discussion will be confined to results from the pretectum, the vestibular nerve and nuclei, and the vestibulo-cerebellum, structures which are relevant for visual-vestibular interaction.

2 Eye Movements

Unblurred vision is mandatory for the orderly processing of visual information during active or passive head movement; image slip on the retina of only a few degrees per second diminishes visual acuity (Westheimer and McKee 1975). Several mechanisms have developed to generate compensatory slow eye movements aided by fast eye movements. Head movements induce compensatory eye movements via the vestibulo-ocular reflex (VOR). Movements of large parts of the visual surround or of single objects induce slow eye movements in the direction of the moving pattern. Continuous rotation of a visual field around the animal induces in both foveate and afoveate species optokinetic nystagmus. In foveate animals movements of a single target object can elicit smooth pursuit eye movements.

Visual-vestibular interactions exist in which eye movements induced by VOR increase image slip instead of reducing it. During head movements within a head-stationary visual surround, or during tracking of a moving object with the head, the slow eye movements elicited by VOR would be directed opposite to the moving scene or object. Such eye movement signals must be cancelled, otherwise they would take the eyes off the target. It is therefore necessary that signals inducing different types of slow eye movements interact within the central nervous system (Dodge 1923; Jung 1948; Cohen 1974). Mowrer (1937) and Ohm (1933) suggested that the interaction of vestibular and visual (optokinetic) informations takes place in the vestibular nuclei. Ito (1970) assumed that the flocculus of the cerebellum might be a further structure to play a crucial role in such interaction because of its multisensory inputs and its output to the vestibular and oculomotor system.

It should be stressed that gaze stabilization is just one of the factors which are important to obtain perceptual stability of the visual world. The subject in all its complexity has been discussed by Steinman et al. (1982).

The motor aspects of nystagmus and eye movement generation in general have been reviewed in several recent symposia (Baker and Berthoz 1977; Kommerell 1978; Cohen 1981; Fuchs and Becker 1981; Lennerstrand et al. 1982; Roucoux and Crommelinck 1982). For rapid eye movement generation, the reticular formation plays an essential role. Lesions of small parts of the reticular formation can lead to permanent loss of rapid eye movement generation and thus loss of fast phases of nystagmus, but otherwise do not seem to affect information processing in the vestibular nuclei (Jaeger et al. 1981; Henn et al. 1984). Thus mechanisms of rapid eye movement generation can be separated from generation of compensatory slow movements. Only the latter will be considered in this review. Visually elicited saccades are discussed in the review by Fischer (this Vol.).

2.1 Measurement of Eye Movements

A method widely used to measure eye movements in animals is to implant silver-silver chloride electrodes connected to DC amplifiers to obtain eye position (Bond and Ho 1970). Resolution is about 1.0 deg. Torsional eye movements cannot be measured. The scleral search coil technique, using an alternating magnetic field to induce an electrical current in a coil attached to the eye, is more accurate and has the further advantage of absolute calibration of eye movements in untrained animals (Robinson 1963; Collewijn 1977; for review Robinson 1981b). For measuring eye movements around the primary position, an infrared reflection method has been improved for contact-free, high accuracy measurements (Bach et al. 1983). The relevant parameter to characterize slow eye movements is their velocity.

It is obtained by differentiating the eye position signal electronically or with a computer. The gain of slow eye movements is measured as their velocity relative to that of the stimulus. For a gain of 1.0, velocities of slow eye movements and of the stimulus are equal.

2.2 The Vestibulo-Ocular Reflex

The three-neuron VOR is composed of primary vestibular neurons, secondary neurons in the vestibular nuclei, and oculomotor neurons (Lorente de Nó 1933). The semicircular canals of the labyrinth act as angular accelerometers and the otolith organs as linear accelerometers. The signals from the canals and the otoliths are conveyed via the primary vestibular neurons in the eighth nerve to central structures where they undergo further processing.

Lorente de Nó (1931, 1933, 1938, 1939) already recognized the functional importance of other, polysynaptic pathways via the cerebellum and reticular formation: "That the vestibular impulses really have at their disposal different anatomic paths is demonstrated by the fact that no labyrinthine excitation necessarily sets up a determinant reaction" (1933).

An adequate stimulus to induce VOR in the horizontal direction is an angular velocity change of the head about the space-vertical axis (yaw rotation) in darkness. The VOR is an open-loop reflex, i.e., its

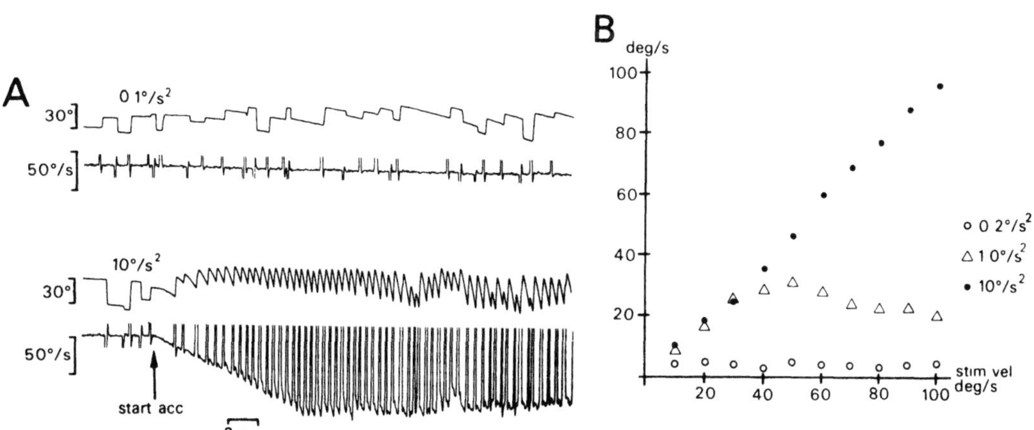

Fig. 1A. Vestibular nystagmus elicited by constant accelerations of $0.1^{\circ} \cdot s^{-2}$ and $10^{\circ} \cdot s^{-2}$ in the monkey. ***First trace,*** horizontal eye position; ***Second trace,*** eye velocity with fast phases clipped. The start of acceleration is indicated by an ***arrow,*** duration of acceleration is 1000 s for $0.1^{\circ} \cdot s^{-2}$ and 10 s for $10^{\circ} \cdot s^{-2}$. **B** Average nystagmus slow phase velocity (***ordinate***) for left and right rotation as a function of instantaneous stimulus velocity (***abscissa***) during three different accelerations (see also Fig. 10a)

performance does not feed back to the labyrinth. It has a short latency in the order of 10–20 ms. The threshold for vestibular nystagmus (VN) in response to an acceleration pulse is below $0.1^{\circ}\cdot s^{-2}$ in monkeys. Nystagmus velocity reaches a gain near unity (0.8–1.0) only for accelerations above $1-10^{\circ}\cdot s^{-2}$ (Fig. 1). The gain is time dependent, that is with prolonged accelerations it progressively diminishes; eye velocity levels off after a certain time and reaches a plateau which is approximately maintained during the remaining part of the acceleration phase (Figs. 1B and 10A; Waespe et al. 1980). The time at which the plateau is reached after the onset of the acceleration pulse varies between monkeys. The gain of the VOR, either in the frequency or time domain is near unity only for accelerations above $10^{\circ}\cdot s^{-2}$ (Fig. 1; Lisberger et al. 1981; Miles and Eighmy 1980; Zee et al. 1981). For frequencies between 0.1 and 1.0 Hz the gain is between 0.98 and 1.03 (Miles and Eighmy 1980). In this range eye movements are compensatory to head movements; for

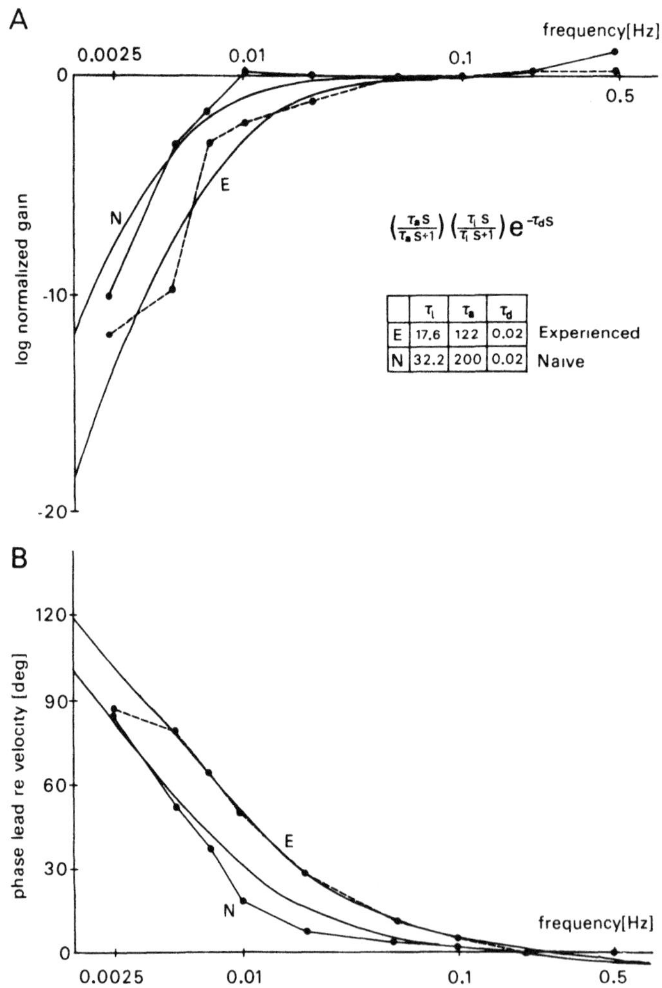

Fig. 2A, B. Gain (**A**) and phase (**B**) of the horizontal VOR during sinusoidal rotation at frequencies between 0.0025 and 0.5 Hz for habituated (eight animals, described as "experienced") and nonhabituated monkeys (five animals, "naive"). Data points are connected by straight line segments with a *solid line* for "naive", and with a *dashed line* for habituated monkeys. *Insert* shows values of model realization represented by *smooth curves* (Buettner et al. 1981)

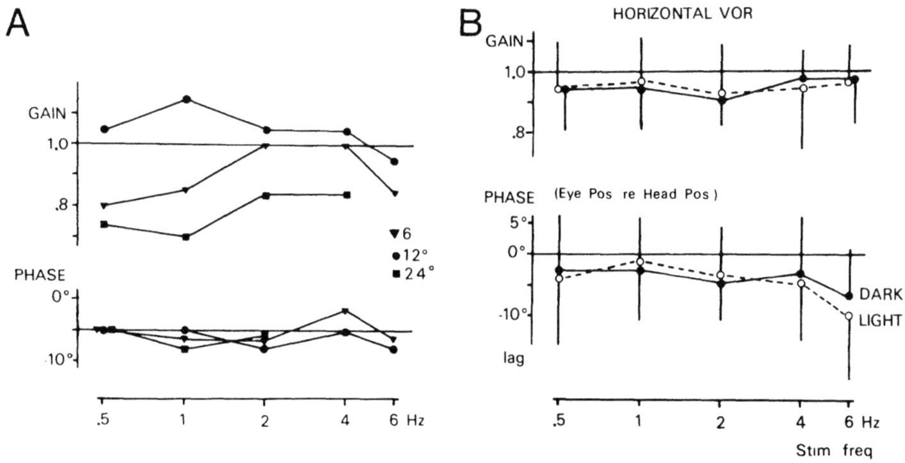

Fig. 3A, B. Gain and phase of horizontal VOR using a sinusoidal velocity profile at frequencies between 0.5 and 6.0 Hz. **A** average values from one monkey using different amplitudes; **B** averaged values for five monkeys (*vertical bar,* 1 SD). *Filled cirles,* rotation in darkness, *open circles,* rotation in the light. Peak acceleration at 4 Hz is $1895^{o} \cdot s^{-2}$ for 6° amplitude, $3790^{o} \cdot s^{-2}$ for 12° amplitude, and $7580^{o} \cdot s^{-2}$ for 24° amplitude (Böhmer and Henn 1983)

an ideal compensation the phase angle between eye and head velocity is defined as 0°. Pseudorandom stimuli gave similar results (Furman et al. 1982). The gain progressively declines with a corresponding increase of the phase angle for frequencies below 0.01 Hz (Fig. 2; Buettner et al. 1981). For frequencies above 1.0 Hz we found no gain increase (Fig. 3; Böhmer and Henn 1983; Paige 1983) in contrast to Keller (1978). In the squirrel monkey the average gain is 0.86 for frequencies between 0.1 and 4 Hz and there is no phase lag (Paige 1983). At 0.2 Hz but not at 0.02 Hz, VOR behaves linearly for velocities between 40 and $360^{o} \cdot s^{-1}$, i.e., gain and phase are independent of stimulus intensity (Paige 1983).

As stated in the introduction (sect. 1) we cannot include a discussion of adaptive changes of the VOR. Nevertheless, it is important to note that all these measurements of gain and phase have been made in normally reared animals. In dark-reared cats, the gain of the VOR is only 50%–65% of normal. There are substantial abnormalities in the phase relationship of head and eye velocity. At 0.3 Hz (± 10° rotation amplitude) a phase lead of 180° was observed, i.e., the eye movements induced by the VOR are anticompensatory (Cynader 1985). One can conclude that the normal development of VOR depends on exposure to visual stimuli and active, unrestricted movements in the early life of an animal.

After an acceleration pulse, VN velocity slowly declines (Fig. 4). The time course of the decline has been approximated by a single exponential for ease of mathematical description. The time constant then indicates

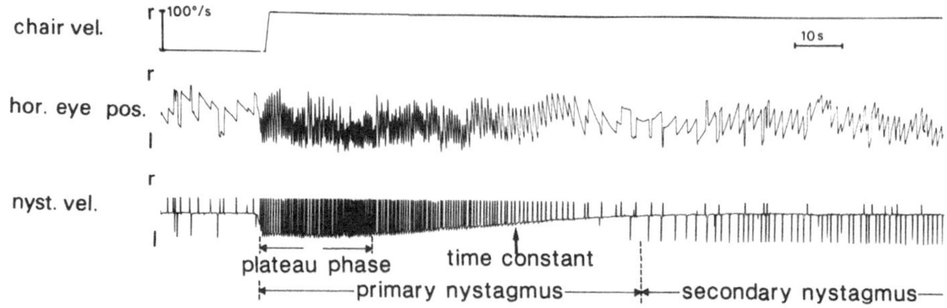

Fig. 4. Horizontal vestibular nystagmus elicited by a velocity step of $100^{\circ}\cdot s^{-1}$ in the dark. *First trace,* turntable velocity; *second trace,* horizontal eye position; *third trace,* horizontal eye velocity. VN velocity remains initially at a plateau for about 22 s before slowly declining. The *arrow* marks the point at which nystagmus velocity has fallen to 37% of its peak initial value; 75 s after the end of acceleration, nystagmus direction reverses and continues as secondary vestibular nystagmus (adapted from Jaeger and Henn 1981)

the time it takes for the eye velocity to decline to 37% of its initial value. The actual decline in eye velocity is, however, more complex. First, after the velocity step, nystagmus velocity does not immediately decline but is approximately constant for several seconds (Fig. 4). This "plateau" in eye velocity is especially pronounced after low velocity steps (Raphan et al. 1979). Secondly, VN can reverse direction after a certain time. This is then defined as secondary VN to differentiate it from primary nystagmus as described above (Fig. 4.). The time constant of primary VN is between 10–40 s and can vary for right and left directions. It depends on the state of habituation (Fig. 2; Sect. 2.5). There is a close correspondence of time constants calculated from the frequency-domain and time-domain responses (Paige1983; Jaeger and Henn 1981).

In summary, VOR induced by changes in angular head velocity in the horizontal plane has a gain near unity for acceleration pulses above $1-10^{\circ}\cdot s^{-2}$ and for frequencies above 0.05 Hz up to the highest frequencies contained in natural head movements. VOR can be considered as a linear system in the range above 0.1 Hz. At low accelerations and for prolonged accelerations, the gain is always below unity and VOR has to be complemented (as during rotation at a constant velocity) by other, mainly visual motion cues.

2.2.1 Horizontal Vestibular Nystagmus as a Function of Head Position

Horizontal VN is strongest when the plane of rotation coincides with the plane of the lateral semicircular canals. When the animal's head is tilted out of this optimal orientation and rotated about an earth-vertical axis, the gain of horizontal nystagmus attenuates with a cosine function. The plane of the lateral canals is tilted about 20° nose-up with respect to the horizontal (Horsley-Clarke) stereotaxic plane with rather large variations between animals (Blanks et al. 1985; Reisine et al. 1985). For an optimal stimulation during earth-vertical rotation, the head of the monkey has therefore to be pitched 20° nose-down in order to make the lateral semicircular canal and rotational planes coplanar.

Horizontal VN can still be elicited, however, after inactivation of the horizontal canals. Inactivation can be achieved by plugging the horizontal canals (Ewald 1892; Money and Scott 1962). After surgically opening the canal, a bone chip is inserted to block the pressure build-up during accelerations. Activity responsible for eliciting horizontal nystagmus can then only arise in vertical canals. Monkeys with both horizontal canals plugged were rotated about the earth-vertical axis while they were positioned in different pitch angles. With the head in the sterotaxic position, horizontal nystagmus reached a gain of 0.4. With the head pitched further down, nystagmus attenuates and reverses if the pitch angle is greater than 33°. This angle then defines the functional null plane for the vertical canals (Böhmer et al. 1985).

2.2.2 Horizontal Vestibular Nystagmus Generated by Off-Vertical Axis Rotation and During Pitching While Rotating

Sustained direction-specific horizontal nystagmus can also be elicited by multidimensional vestibular stimuli which dynamically activate several parts of the labyrinth (for review, Raphan and Cohen 1985).

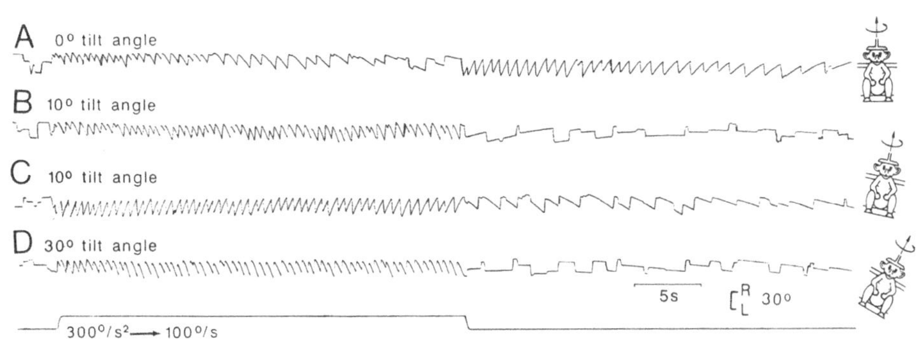

Fig. 5 A–D. Horizontal nystagmus during off-vertical axis rotation (OVAR). Acceleration $300^{o}\cdot s^{-2}$, constant velocity $100^{o}\cdot s^{-1}$. In **A**, during rotation about the earth-vertical axis nystagmus velocity slowly declines after the velocity step to zero. Deceleration induces nystagmus in the opposite direction. In **B–D**, with varying off-vertical angles, nystagmus is continuous during rotation at a constant velocity and there is little nystagmus after the deceleration, In **A**, **B**, **D** rotation to the right, in **C** to the left (Henn et al. 1983)

When an animal is rotated about the earth-vertical axis, the labyrinths do not change their position relative to gravity, and perrotatory nystagmus declines to zero (Figs. 4 and 5A). When the animal, however, is rotated about an axis tilted from the earth-vertical, perrotatory nystagmus is maintained for as long as the stimulus continues (Fig. 5 B–D; Young and Henn 1975; Raphan et al. 1981). This continous horizontal nystagmus during off-vertical axis rotation (OVAR) is direction specific, i.e., rotation to the left causes slow phase eye velocity to the right, and vice versa during rotation to the right. It is present with tilt angles of only 5–10°. Steady-state velocities of nystagmus increase up to stimulus velocities of 60–90$^{o} \cdot s^{-1}$ and saturate with eye velocities between 40–60$^{o} \cdot s^{-1}$ (Fig. 6). At higher velocities nystagmus velocities can decline again.

During OVAR the labyrinths continuously change their position relative to gravity. The rotating gravity vector then activates the otolith afferents and generates positional information about the attitude of the head with regard to gravity (Guedry 1965). This positional information needs further central processing in order to be converted into a velocity command signal. Raphan et al. (1981) have proposed that this velocity command signal is formed centrally by detecting a moving pattern which is induced by the sequential activation of otolith afferents with different polarization vectors.

Although tilts during per-rotatory nystagmus cause nystagmus velocity to be maintained, tilts during postrotatory nystagmus cause a rapid decline in slow phase velocity and VN is cancelled within 5–10 s. This phenomenon has been called "tilt-dump" of postrotatory nystagmus (Raphan and Cohen 1985; humans, Schrader et al. 1985). Bilateral macular ablation strongly reduces this "tilt-dump" of post-rotatory VN in the squirrel monkey (Igarashi et al. 1980).

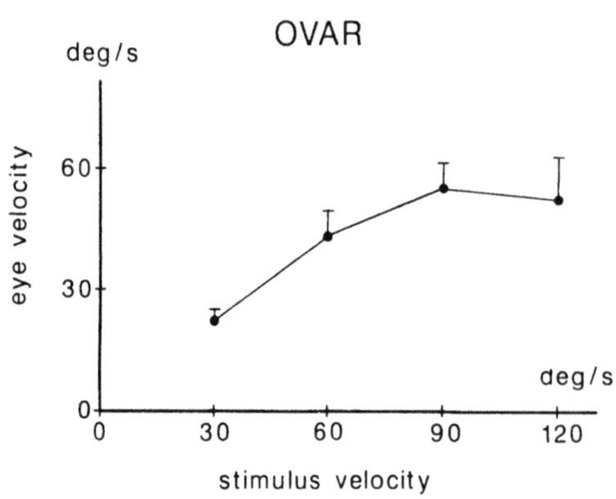

Fig. 6. Steady-state slow phase velocity of horizontal nystagmus during off-vertical axis rotation (OVAR) for different constant velocities of rotation *(abscissa)*. The rotation axis is tilted 30° off the earth-vertical axis. Averages of six monkeys, *vertical bars* indicate 1 SD

Sinusoidal pitch about an earth-horizontal axis in darkness induces sinusoidal vertical nystagmus. If the animal, in addition to the pitching, is rotated about an earth-vertical axis at constant velocity, a sustained horizontal nystagmus is also induced which persists for the duration of stimulation (Raphan et al. 1983a; review, Raphan and Cohen 1985). Steady-state horizontal nystagmus reaches peak velocities of 30–50$^{\circ} \cdot s^{-1}$ and is independent of pitch periods between 5–20 s. Cessation of pitch during continued rotation is followed by a slow decline of horizontal slow phase velocity with a time constant of 10–20 s. Sinusoidal roll (rotation about the naso-occipital axis) while rotating elicits much less or no horizontal nystagmus. By selectively plugging different canal pairs, it was shown that activation of the vertical canal afferents plays the predominant role in generating the continuous horizontal nystagmus (Raphan et al. 1983a).

2.2.3 Conclusion

Horizontal vestibular nystagmus can also be induced by activation of the vertical canals or the otoliths if certain patterns of stimulation are applied. Under natural conditions in a freely moving animal, most movements would simultaneously excite all canals and the otoliths. Collewijn (1977) found the gain of eye movements in darkness to be unity in freely moving rabbits, while it could be considerably lower if only one canal pair was stimulated in isolation. This shows that not only visual input, as discussed in the next section, but vestibular input arising in different labyrinthine receptors can combine to give rise to an eye velocity signal which can match head velocity rather accurately. Proprioceptive inputs from the neck do not contribute to VOR in a normal monkey (Dichgans et al. 1973).

2.3 Visually Induced Slow Eye Movements

Slow following eye movements are induced by movement of the entire visual surround or by movement of a small target object. Continuous rotation of the visual surround elicits optokinetic nystagmus (OKN). Foveation of a small target light induces smooth pursuit (SP) eye movements. As a target usually moves in front of a stationary background, different motion signals are induced by the central and peripheral retina. If a single target is tracked by the fovea and the eye moves across a stationary background, the peripheral retina is stimulated with a continuous motion signal. In the absence of a foveal target, this peripheral motion signal would be a powerful stimulus to induce optokinetic nystagmus which has to be overriden in the presence of a foveal target. It is therefore important to determine whether the dynamics of OKN and SP are similar and whether they have pathways and mechanisms in common or not.

2.3.1 Optokinetic Nystagmus

The nystagmus induced by an optokinetic or vestibular stimulus is phenomenologically the same. OKN is composed of slow following eye movements (slow phase) in the direction of the stimulus, and fast resetting saccades (fast phase) in the opposite direction. A thorough historical overview of OKN has been given by Bender and Shanzer (1983). OKN is best elicited by movement of a full field which stimulates the entire retina. Stimulation of the fovea or parafovea is not essential for eliciting OKN. Nevertheless, the strongest responses in humans and monkeys are induced by stimulation of the foveal region (Koerner and Schiller 1972; Dubois and Collewijn 1979; van Die and Collewijn 1982). No asymmetry in monocular optokinetic nystagmus was found for nasotemporal and temporonasal stimulation in monkeys or humans. Such an asymmetry was demonstrated in animals without a fovea (Tauber and Atkin 1968). In human infants also monocular stimulation in the temporonasal direction elicits higher velocity nystagmus than stimulation in the nasotemporal direction during the first few months of life while the fovea is still immature (Atkinson 1979; Naegele and Held 1982).

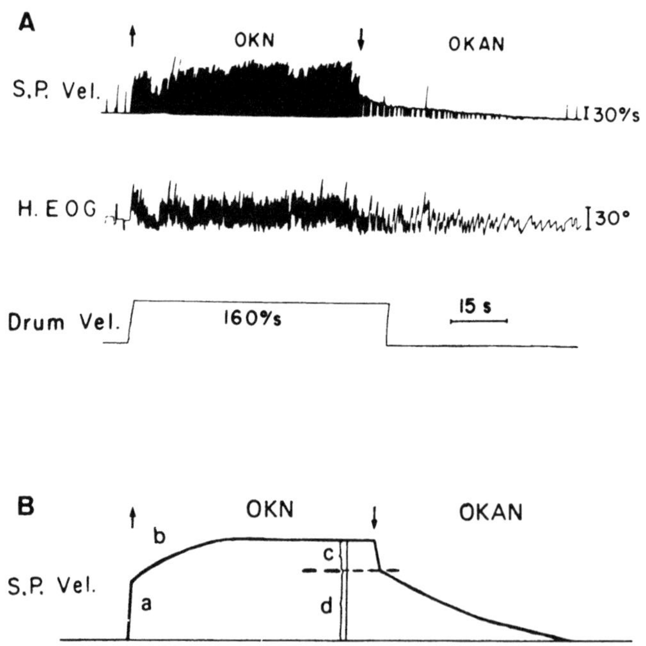

Fig 7A. OKN and OKAN in response to a velocity step of the optokinetic drum. *First trace,* velocity of horizontal eye movements, fast phases clipped; *second trace,* horizontal eye position; *third trace,* velocity profile of the optokinetic drum. *Upward arrow,* lights on; *downward arrow,* lights out. **B** Velocity envelope of the slow phases of OKN and OKAN redrawn schematically from the first trace in **A**

This asymmetry can persist into adult life with congenital or early acquired amblyopia.

The slow phase velocity of OKN in response to a step in stimulus velocity has two dynamically different components (Fig. 7). A rapid initial change (a) in velocity within the first second is followed by a slower rise (b) to a steady-state level (Cohen et al. 1977; Paige 1983). When the light is turned out, eye velocity rapidly drops (c) to a value (d) from which it slowly decays as optokinetic after-nystagmus (OKAN). Cohen et al. (1977) have proposed that two different "mechanisms" are responsible for these components (Fig. 12). "Direct" visual-oculomotor pathways determine the fast dynamics (a, c) and "indirect" pathways the slow dynamics (b, d). Essential in the "indirect" pathways is a "velocity storage element". Robinson (1981a) used a different connotation with the "smooth pursuit" system which is responsible for the fast dynamics, and the "optokinetic" system which corresponds to the "indirect" pathways.

The eye velocity reached by the initial fast jump of OKN in monkeys (a in Fig. 7) is 40%–80% of the stimulus velocity and progressively decreases with greater stimulus velocities (Fig. 8B; Cohen et al. 1977;

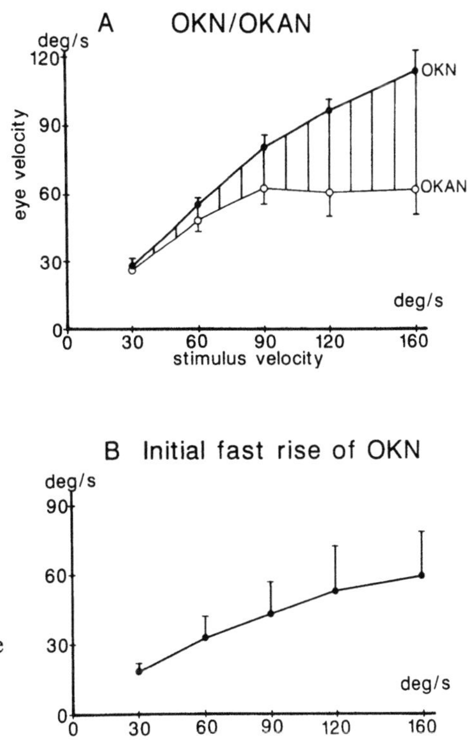

Fig. 8A. Steady-state slow phase velocity of OKN (*filled circles*) and initial peak OKAN velocities (*open circles*) as a function of constant stimulus velocity (*abscissa*). *Hatched area* indicates fast decay in eye velocity in the transition of OKN to OKAN. **B** Initial fast increase in slow phase velocity of OKN at the onset of stimulation. Averages of six monkeys, *vertical bar* 1 SD

Paige 1983). The time constant of the slow increase in eye velocity (b in Fig. 7) gets larger with higher stimulus velocities. Steady-state values of the slow phase velocity are almost equal to stimulus velocities up to $60^{\circ}\cdot s^{-1}$ (Fig. 8A; Paige 1983; Zee et al. 1981; Lisberger et al. 1981). For higher velocities, gain progressively declines; at $120^{\circ}\cdot s^{-1}$ stimulus velocity, average gain is usually between 0.8 and 0.9 (Paige 1983; Waespe et al. 1983). OKN is more irregular at these higher velocities. Therefore, average gain is always smaller than peak values maintained only for short time periods. The amount of the fast decline in eye velocity in the transition from OKN to OKAN (c in Fig. 7) is minimal up to stimulus velocities of $60^{\circ}\cdot s^{-1}$ (Fig. 8A; Cohen et al. 1977). With increasing velocities it also increases. In monkeys with central scotomata, the dynamic characteristics of OKN are not changed qualitatively (Büttner et al. 1983). Even with central scotomata of 12° (fovea diameter: 6°) the effects on the initial fast jump and on steady state values are negligible. With central lesions of more than 30°, the amount of the initial jump is reduced by about 50%, but eye velocities of up to $40^{\circ}\cdot s^{-1}$ are still elicited. In humans, a central region of 5° is as powerful in eliciting OKN as the entire retina peripheral to this area (Collewijn et al. 1982).

In response to sudden illumination of the moving full field, the latency of slow eye movements is between 100–150 ms (Lisberger et al. 1981; Waespe et al. 1985b). Using a random dot pattern and a field restricted to the central region of the retina, a shorter latency of around 50 ms was found by Miles et al. (1986). The authors discuss several reasons why these shorter latencies have not been observed in previous studies. The initial eye acceleration within the first 100 ms after onset of the slow eye movement is between 200 and $400^{\circ}\cdot s^{-2}$ and depends on the magnitude of the stimulus velocity (Lisberger et al. 1981). Applying acceleration steps, OKN velocity has a gain near unity, that is, it follows closely the instantaneous velocity of the full field for accelerations up to $5^{\circ}\cdot s^{-2}$ and velocities of $100^{\circ}\cdot s^{-1}$ (Waespe et al. 1980, Fig. 10B). Gain rapidly falls with accelerations above $10^{\circ}\cdot s^{-2}$. The OKN is determined by the response limitations of the system not only to the velocity but also to the acceleration of the stimulus.

Under open-loop conditions (i.e., the optokinetic stimulus is presented to a paralytic eye and eye movements are recorded from the nonparalytic, covered eye) the highest gain is 10–30 at low stimulus velocities of $2\text{-}5^{\circ}\cdot s^{-1}$ (Koerner and Schiller 1972). At a stimulus velocity of $30\text{-}40^{\circ}\cdot s^{-1}$, gain is near unity and declines for higher stimulus velocities. In humans, Dubois and Collewijn (1979) found similar values.

Applying sinusoidal (predictable) optokinetic stimulation, the gain of OKN is between 0.8 and 0.95 for frequencies up to 0.2 Hz (peak velocity $40^{\circ}\cdot s^{-1}$). At 0.01 Hz it is between 0.93 and 0.95 with eye and stimulus velocities being in phase (Paige 1983; Boyle et al. 1985). A substantial OKN can still be elicited at frequencies of 2–4 Hz with gains between 0.3–0.5, but eye velocity relative to stimulus velocity lags 40°–100° (Paige 1983; Boyle et al. 1985). In dark-reared cats OKN can be induced immediately after the first exposure to optokinetic stimulation but with a lower gain as compared to normally reared cats. This is especially pronounced at high stimulus velocities (Cynader 1985). Cats which have been raised in an 8 Hz stroboscopically illuminated environment (flash duration 10 μs), depriving them of experience with visual movement but not with form vision, had virtually no direction selective neurons in the visual cortex (Cynader and Chernenko 1976), but smooth pursuit and binocular OKN were comparable to normal cats (illumination rate 8 Hz, duration 4 μs flash). With monocular stimulation, however, OKN was virtually absent for stimulation in the naso-temporal direction (Conway et al. 1981). Continuous image motion on the retina is obviously not mandatory for the development of visually induced slow eye movements. In cats raised in an 0.7 Hz stroboscopically illuminated environment, visually induced slow eye movements were abnormal and no smooth pursuit could be elicited.

In summary, visual control of eye movements is restricted by a long delay and limitation in response to high stimulus velocities and accelerations. The optokinetic system works well for velocities up to 60°–$90^{\circ}\cdot s^{-1}$ and accelerations up to $5^{\circ}\cdot s^{-2}$. In contrast to the VOR, the optokinetic system operates as a negative feedback system with a high open-loop gain. Sinusoidal stimulation, i.e., examination in the frequency domain, does not reveal the two, dynamically different components in the primate OKN slow phase velocity response. Qualitatively, OKN is the same in monkeys and humans, quantitatively there are differences, especially in the dynamics of the slow component (Cohen et al. 1981).

2.3.2 Optokinetic After-Nystagmus

Primary optokinetic after-nystagmus (OKAN) occurs in darkness following OKN and beats in the same direction. In the transition of OKN to OKAN, eye velocity rapidly drops within 1 s to values from which it further declines over a much longer time course. As noted above, in the monkey the amount of the initial fast decline is minimal up to OKN velocities of 40°–$60^\circ \cdot s^{-1}$. With higher OKN velocities the OKAN saturates, leading to an increasing velocity drop from OKN to OKAN (Fig. 8A; Cohen et al. 1977). The maximal initial velocity of OKAN, which is called its saturation velocity, is usually 50°–$70^\circ \cdot s^{-1}$ (Fig. 8A; Lisberger et al. 1981; Waespe et al. 1983), but values up to $120^\circ \cdot s^{-1}$ have been reported (Cohen et al. 1977). In humans, this value is much smaller not exceeding $20^\circ \cdot s^{-1}$ (Cohen et al. 1981; Koenig and Dichgans 1981; Segal and Liben 1985). Primary OKAN velocity declines over a time course similar to VN velocity and can be characterized by a time constant of 10–40 s. Time constants of primary VN and OKAN closely correspond to each other (Raphan et al. 1979). After variable periods, primary OKAN is followed by secondary OKAN which beats in the opposite direction. Secondary OKAN itself can eventually be followed by tertiary OKAN which beats again in the direction of primary OKAN (Koerner and Schiller 1972; Waespe and Henn 1978a; Waespe et al. 1978). When the visual full field is decelerated in the light, OKAN velocity closely follows the decreasing stimulus velocity for decelerations up to $10^\circ \cdot s^{-2}$. For higher decelerations, nystagmus outlasts the end of deceleration as OKAN (Fig. 10E; Waespe et al. 1980). When the full field is suddenly stopped in the light, OKAN is short and suppressed within 4–6 s (Cohen et al. 1977).

2.3.3 Sigma-Movement

When slow eye movements are induced across a periodic visual pattern (e.g., a grid of black and white stripes or a row of equally spaced dots) which is illuminated stroboscopically, subjects can perceive an illusory movement of the stationary pattern in the direction of the slow eye movements (Behrens and Grüsser 1982). This illusory movement then maintains the slow eye movements which in turn produce the illusion. The critical parameter for the illusion is that the eyes move over the distance of one or more periods of the pattern between two flashes of the stroboscopic illumination. The phenomenon has been called sigma-movement and sigma-OKN if nystagmus is elicited. Sigma-OKN can also be induced in monkeys when the stroboscopically illuminated periodic visual pattern is presented during postrotatory nystagmus or OKAN (Grüsser et al. 1979). Sigma-OKN can reach velocities of up to $90^\circ \cdot s^{-1}$ and is followed in darkness by sigma-OKAN (Adler et al. 1981). The angular velocity of the sigma-OKN increases with the period of the stationary stimulus pattern and the flash frequency. Sigma movement is one of the many examples which demonstrate that the stimulus for a slow eye movement is the perception of movement rather than an actual movement of an object across the retina. Other such phenomena have been discussed by Young (1977).

2.3.4 Smooth Pursuit Eye Movements

Smooth pursuit eye movements (SP) are defined as following movements in foveate animals elicited by a single moving target. SP is usually initiated by the motion of an object across the retina and is guided by visual feedback. Determining factors are the object's velocity and eccentricity relative to the fovea (Rashbass 1961; Robinson 1965; Kommerell and Klein 1971; Fuchs 1967). Parafoveal smooth pursuit is easily accomplished without practice (Winterson and Steinman 1978). Recently Lisberger et al. (1981) and Lisberger and Westbrook (1985) have suggested that the determining factor to guide SP is not the velocity but the acceleration of the target, and that the visual system provides an eye acceleration command to the SP system. The reaction time of the SP system has been measured in monkeys to be 75 ms (Lisberger and Westbrook 1985) or 175 ms (Fuchs 1967), and 125 ms in humans (Robinson 1965). Peak velocities exceed $100^{o} \cdot s^{-1}$ in monkeys (Lisberger and Fuchs 1978a; Barmack 1970) and reaches $90^{o} \cdot s^{-1}$ on average in humans (Meyer et al. 1985). The gain of SP in humans induced by tracking a predictive periodic target falls rapidly at frequencies above 1.0 Hz (Barnes et al. 1978). Gain is not only a function of target velocity, but also of target acceleration (Lisberger et al. 1981), similar to OKN (Waespe et al. 1980). Collewijn and Taminga (1984) have meticulously analysed SP in humans. Even normals pursue with many catch-up saccades in the direction of the moving target. At 0.5 Hz ($\pm$ 10°) the gain of the smooth part of the tracking eye movements (gain measured as the amplitude ratio of eye and target position) for a sinusoidally oscillating target was only 0.66. Catch-up saccades increase, however, the gain of the tracking eye movements to almost unity.

SP is not followed by a substantial after-response in monkeys (Lisberger et al. 1981), but in humans SP after-responses can reach initial velocities of up to $10^{o} \cdot s^{-1}$ (Muratore and Zee 1979). SP eye movements can also be induced by a stabilized retinal image eccentric to the fovea (Kommerell and Täumer 1972; Pola and Wyatt 1980). Steinbach (1976) suggested that the stimulus for pursuit eye movements may be the perception of an object in motion regardless of the retinal stimulation. Subjects, for instance, succesfully track the invisible center of a rotating wheel when the retinal stimuli are points of light moving in a cycloid on the rim of the rotating wheel, even when no actual motion has occurred in the direction of the eye movements.

SP eye movements induced by foveation of a single object are normally made in light. Under these circumstances, a stationary background could be an effective optokinetic stimulus for slow eye movements in the opposite direction as soon as the target is tracked. Collewijn et al. (1982) found that the velocity of SP in man is diminished by 20%–30% in the presence of a stationary background. The deficit is supplemented by catch-up saccades and the overall accuracy of pursuit was unaffected. In another publication Collewijn and coworkers concluded that any observed influence of a moving or stationary background on smooth eye movements can be attributed to the subjects' effort or attention rather than to the operation of an involuntary mechanism that integrates velocity information from target and background (Kowler et al. 1984). Man and monkey thus can easily perform SP in light and this points towards a mechanism whereby foveal motion information can override motion information from the peripheral retina.

If humans or monkeys fixate on a stationary target in front of a moving full field, the eyes are not held stationary on the target, but intermittently move in the direction opposite to the moving full field (Collewijn et al. 1982). Humans then perceive the target to move in the same direction as the eyes and opposite to the full field. Under these conditions humans pursue an imaginary target motion opposite to the moving field.

In summary, SP eye movements are elicited not only by foveal, but also by extrafoveal stimulation. Visual velocity error feedback is most important in eliciting SP eye movements, although both retinal position errors and perception of target velocity may also make a significant contribution. The velocity gain is near unity for low accelerations and low velocities, but even under normal conditions smooth eye movements are supplemented by catch-up saccades. As soon as the eyes track a target across a structured background an optokinetic stimulus is also presented, but in the opposite direction. Velocity of SP eye movement is, however, only slightly diminished by the optokinetic stimulus. OKN and SP are probably governed by separate but partially overlapping subsystems. High accelerations or velocities of the optokinetic response are assumed to be generated by the SP system (Robinson 1981a; Waespe et al. 1985b). The SP system probably operates differently in response to various types of stimuli, also depending on the mental set of the subjects, i.e., whether the visual target to be followed is of interest or not. Undoubtedly, cerebral cortical mechanisms are involved in the decision, volitional or not, as to which target should be followed and which retinal motion information should be neglected.

2.4 VOR-Gain Changes by Nonvisual Cues

Yasui and Young (1975) demonstrated that tracking of a foveal afterimage during angular head oscillations in the dark elicits eye velocitites which exceed those for normal vestibular nystagmus. A perceived target (afterimage) motion was used to generate eye movement commands and to increase the VOR-gain.

Similarly, fixating an imagined space-stationary or subject-stationary target during passive predictable oscillation (at 0.3 Hz) in darkness increases the VOR-gain to 0.95 or decreases it to 0.35, respectively (Barr et al. 1976). Even when the passive head oscillations were unpredictable, the same effects could be observed (McKinley and Peterson 1985). As SP eye movements are much better for predictable stimuli, the authors concluded that mechanisms which control SP and voluntary control of VOR are not necessarily identical.

The gain of the VOR is also dependent on target location for pure geometrical reasons. For close targets the VOR-gain has to be adjusted considering four parameters: the position of the subject relative to the center of rotation, the distance of the target from the center of rotation, the interocular distance, and the target eccentricity. Viirre et al. (1986) demonstrated in monkeys that VOR-gain increases with increasing eccentricity of head rotation and with decreasing target distance. The amount of VOR-gain increase agreed closely with predictions made on the basis of geometrical relationships. The VOR-gains of the two eyes were also different because the distance between each eye and the target was different. These results are important because in the usual laboratory situation the target distance is about 50 cm and the eyes are 5 cm eccentric to the rotation axis; VOR-gain is then above unity. The effects on VOR-gain increase could still be observed at 2.0 Hz, when visual influences become less effective. The authors conclude that the gain increase due to larger eccentricity of the subject is effected by otolith-canal interactions, and that visual information about target distance and accomodation "presets" the VOR-gain to higher values.

The VOR-gain can also be modified by changed otolith input. During parabolic flight horizontal nystagmus velocity is enhanced with increased gravity and reduced during zero gravity. The ratio of nystagmus velocity at 1.8 G and 0 G was 1.7 (Lackner and Graybiel 1981).

2.5 Habituation

If a monkey is repeatedly exposed to vestibular stimuli, VN shortens. Velocity steps or low frequency sinusoidal stimulation are effective in producing habituation (Jaeger and Henn 1981). The shortening of the dominant decaying time constant of VN is manifested in the frequency domain by an increasing phase advance of velocity of the VN waveform relative to that of the stimulus (Fig. 2). Whereas time constants of VN can be 30–60 s in experimentally naive monkeys they reduce to 6–15 s in habituated monkeys. Gain of VN velocity also decreases during low frequency stimulation (Jaeger and Henn 1981). In response to velocity steps the VN time constant is similarly reduced, but the gain of the step response does not change. In habituated monkeys the duration of the plateau phase is reduced and VN starts to decline immediately after the end of the velocity step. Again, humans show qualitatively the same behavior (Baloh et al. 1982).

The time constant of primary OKAN also rapidly shortens (to 10–20 s) in response to repetitive optokinetic stimulation (Cohen et al. 1977; Waespe and Henn 1978a). Secondary OKAN, however, always increases in amplitude and duration (Waespe and Henn 1978a). In contrast, secondary VN does not increase but usually also decreases like primary VN with repetitive vestibular stimulation (Waespe, unpublished). The functional role of these different effects of habituation on VN and OKAN has not been clarified.

These considerations show that the response of the vestibular system to the same stimulus can vary over wide ranges in the normal human subject and monkey. The response depends on the recent history (habituation), interaction with otolith input, and the mental set of the subject.

2.6 Visual-Vestibular Interaction

The dynamics of neither the vestibular nor the optokinetic system alone sufficiently cover the full spectrum of movements required to generate compensatory eye movements for all accelerations and velocities. The optokinetic system works well at low accelerations and low constant velocities and the vestibular system at high accelerations.

2.6.1 Rotation in the Light

The vestibular and optokinetic reflexes function together when the monkey is rotated in a space-stationary environment (Fig. 9). For sinusoidal stimulation the gain of the combined nystagmus has an average of 0.96 with no phase shift between nystagmus and stimulus velocities for all frequencies tested (0.01–4.0 Hz, Paige 1983). For acceleration steps the gain is also above 0.9 for accelerations between 2.5 and $40^{\circ} \cdot s^{-2}$ regardless of the duration of the acceleration up to instantaneous velocities of $160^{\circ} \cdot s^{-1}$ (Fig. 10C; Waespe et al. 1980).

Eye velocity is maintained during constant velocity rotation with a gain similar to that during constant velocity optokinetic stimulation (Fig. 9B, C). For rotations above $120^{\circ} \cdot s^{-1}$ eye velocity therefore slowly

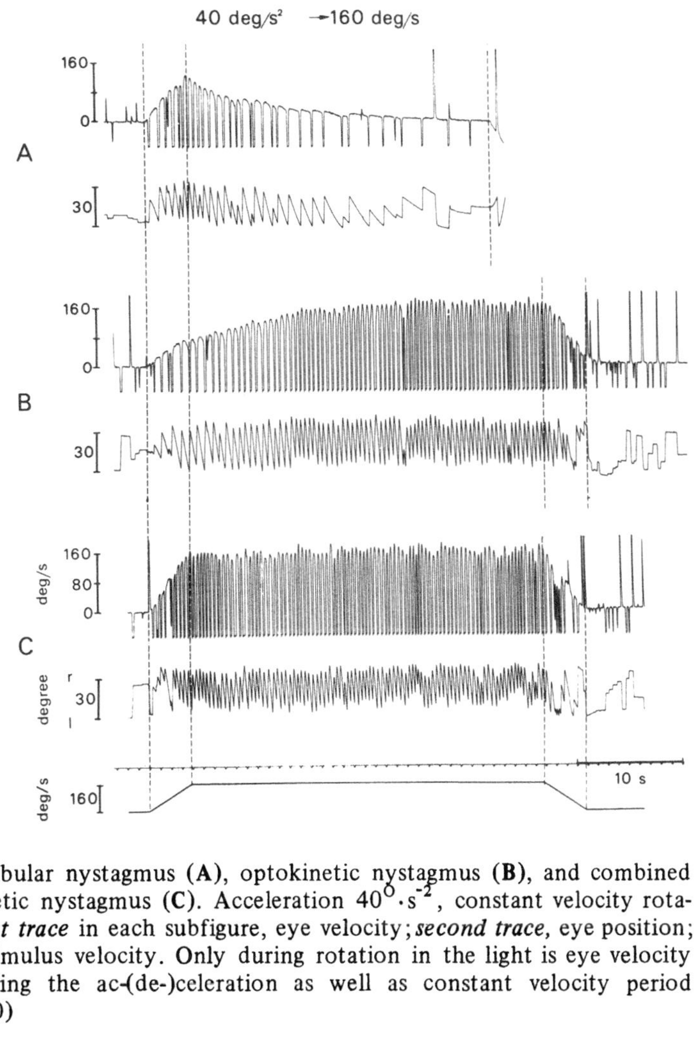

Fig. 9 A–C. Vestibular nystagmus (**A**), optokinetic nystagmus (**B**), and combined vestibular-optokinetic nystagmus (**C**). Acceleration $40^{\circ} \cdot s^{-2}$, constant velocity rotation $160^{\circ} \cdot s^{-1}$. *First trace* in each subfigure, eye velocity; *second trace,* eye position; *last trace* in **C**, stimulus velocity. Only during rotation in the light is eye velocity compensatory during the ac-(de-)celeration as well as constant velocity period (Waespe et al. 1980)

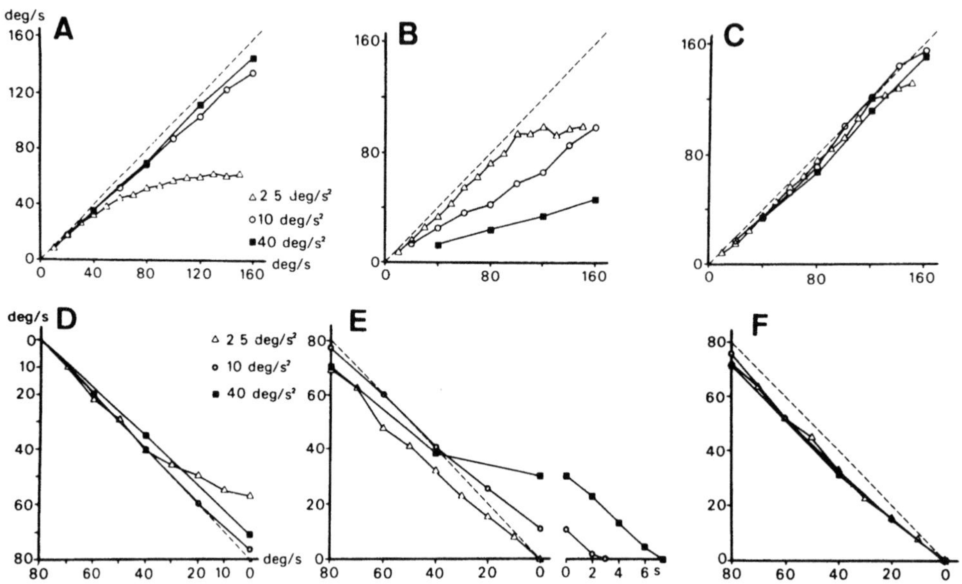

Fig. 10 A–F. Slow phase velocity of vestibular nystagmus (**A, D**), optokinetic nystagmus (**B, E**), and of combined nystagmus (**C, F**) as a function of instantaneous stimulus velocity (*abscissa*) which increased (**A–C**) or decreased (**D–F**) at a rate of 2.5, 10, and $40^{\circ} \cdot s^{-2}$. Averages from one monkey. The *dashed line* indicates a gain of unity. The turntable (**D, F**) or the visual surround (**E**) are decelerated from a constant velocity of $80^{\circ} \cdot s^{-1}$. In **E** after the end of deceleration of the visual surround, a time scale is added, as OKN outlasts the end of the deceleration by several seconds (Waespe et al. 1980)

decays after the end of acceleration as the OKN gain in this velocity range is smaller than the gain of VOR (Cohen et al. 1977). During deceleration of the monkey in the light nystagmus velocity closely follows the instantaneous decreasing turntable velocity (Fig. 10F). Cohen and coworkers (Cohen et al. 1977; Raphan et al. 1979) further analyzed the interaction of OKAN and VN. When after continuous rotation of the monkey in the light the turntable is suddenly stopped and the light turned off, a postrotatory VN is not observed as long as the eye velocity during the preceding rotation did not exceed the saturation velocity of OKAN. OKAN induced during rotation in the light and postrotatory VN induced by the sudden stop of the animal in darkness summate and cancel each other.

2.6.2 Conflict Stimulation and Suppression of VOR and OKAN

In certain conditions eye movements induced by head movements must be suppressed to maintain image stability on the retina. One example is tracking of a target with the head. VOR, if not suppressed, would then move the fovea off the target Suppression of VN by a subject-stationary full field is generally called conflict stimulation, and suppression of VN by a single target VOR-suppression. The difference is that during conflict stimulation large parts of the retina and the fovea are stimulated regardless of eye position, whereas during VOR-suppression only the foveal and parafoveal region are alternatively activated as long as VOR-suppression is not complete.

Acceleration of the monkey together with the animal-stationary full field surround in the same direction strongly reduces VN. For accelerations up to $10^{o} \cdot s^{-2}$ nystagmus is almost completely suppressed and peak velocity is less than 10% of the nystagmus velocity in darkness. In response to a velocity step, peak nystagmus velocity is around 25%–40% during conflict stimulation compared to the velocity in darkness. The time constant of VN is always short and reduced to 4–6 s during conflict stimulation (Waespe and Henn 1978b; Büttner et al. 1983).

Cohen and coworkers (Cohen et al. 1977; Raphan et al. 1979) used another paradigm to measure fixation suppression of VN and OKAN. During postrotatory VN or OKAN they turned lights on for short periods, thereby exposing the monkey to an animal-stationary full field (Fig. 11). When lights were turned off again, nystagmus resumed at a lower velocity than would have occurred without fixation. This paradigm measures how fast the velocity storage mechanism responsible for VN and OKAN is discharged during the period of fixation. Because nystagmus velocity after fixation is lower than without fixation the storage mechanism must have lost activity during the short period of full field fixation. This has been called "dumping". The time constant is strongly reduced to ranges between 1.5 and 4.8 s during VN and bet-

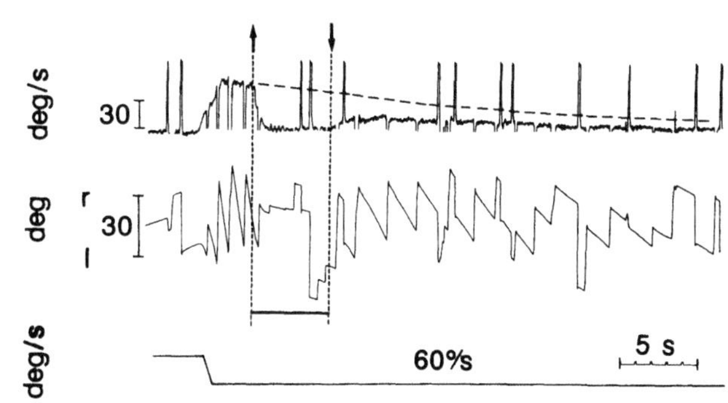

Fig. 11. Effects of fixation of a monkey-stationary full field on slow phase velocity of vestibular nystagmus and its recovery. The traces are eye velocity, eye position, and turntable velocity (acceleration $100^{o} \cdot s^{-2}$, constant velocity $60^{o} \cdot s^{-1}$). The period of fixation is indicated by *vertical dashed lines.* Before and after the fixation period the monkey is in darkness. The *dotted line* in the velocity trace is the velocity of vestibular nystagmus without fixation, i.e., monkey is continuously in darkness

ween 1 and 3 s during OKAN (Cohen et al. 1977; Raphan et al. 1979; Waespe and Schwarz 1986). Fixation of a single light instead of a full field also rapidly attenuates activity of the storage mechanism (Waespe and Schwarz 1986; see, however, Lisberger et al. 1981). The time constants are reduced to 4.5 s during VN and 4.0 s during OKAN.

By fixating a single light point monkeys suppress the VOR during sinusoidal passive rotation almost completely up to 1.0 Hz. For higher frequencies the performance rapidly declines and at 2.0 Hz ($\pm 10^{o}$) only 45% of the VOR can be suppressed (Lisberger et al. 1981). The two curves of VOR-suppression and of SP in the frequency domain are very similar in humans and monkeys, i.e., there is almost complete suppression of the VOR when the SP gain is near unity. This led to the hypothesis that a similar mechanism is responsible for the generation of SP and for suppression of VOR. Cancellation of the VOR is assumed to be effected by summation of a pursuit and a VOR-signal with opposite signs (Barnes et al. 1978). This hypothesis has recently been challenged (Robinson 1982; Tomlinson and Robinson 1984).

2.7 Models

It is almost impossible to summarize the wealth of data on visual-vestibular interaction without referring to hypotheses and mathematical models which have inspired many experimental approaches (Robinson 1981b). The main concepts underlying these models have been summarized in Henn et al. (1980). These concepts have not changed for recent years and we would like to refer for details to this and other recent publications (Buizza and Schmid 1982; Raphan and Cohen 1985; Robinson 1981a, b; Galiana 1986). These models can simulate the dominant aspects of visual-vestibular interaction as reflected in single cell recordings and eye movement responses, but it is not yet known how the elements are actually realized by the nervous system.

In the following we discuss two aspects of two of these models which are essential to an understanding of single cell recordings and lesion studies (Robinson 1981b; Raphan and Cohen in Henn et al. 1980). The first important aspect is the separation of two "pathways" by which visual motion information induces slow eye movements during OKN (Fig. 12). This is an idea going back to Ter Braak (1936; translated in Collewijn 1981) who distinguished a "stare nystagmus" (Stiernystagmus) and a "look nystagmus" (Schaunystagmus) to account for the fact that actively pursuing instead of staring at the optokinetic stripes immediately increases OKN velocity. He proposed the idea that

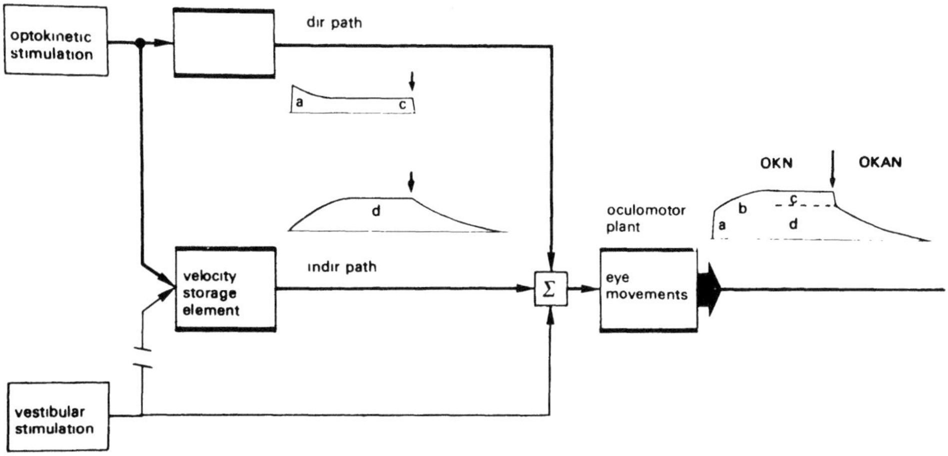

Fig. 12. Scheme of "direct" and "indirect" pathways and their convergence

the "stare nystagmus" is mediated by subcortical pathways, whereas the "look nystagmus" involves cortical structures. These "pathways" are not characterized anatomically as being oligo- or polysynaptic but are rather operationally defined by how rapidly activity builds up in response to image slip velocity to induce slow eye movements. Fast changes in slow phase velocity of OKN (a, c in Figs. 7 and 12) are attributed to activation of "direct" visual-oculomotor pathways which are characterized by their fast dynamics (Cohen et al. 1977; Raphan and Cohen 1985). Robinson (1981b) attributed the generation of these fast components to the smooth pursuit system. In the Raphan-Cohen model the problem of foveal pursuit is not addressed. Slow changes in eye velocity and low steady-state velocities of OKN and the occurrence of OKAN (b, d in Figs. 7 and 12) are attributed to activation of "indirect" pathways. These pathways are identical to the "optokinetic system" of Robinson (1981a, b). The key element of the "indirect" pathways is a mechanism by which information is stored during stimulation and which slowly dissipates its activity. In the Raphan-Cohen model this storage mechanism is represented by a leaky eye velocity integrator with a charcteristic time constant. This storage mechanism is supposed to be common to the vestibular and optokinetic system (see also Demer and Robinson 1983). As the activity of the eye velocity storage mechanism is responsible for the prolongation of VN and the occurrence of OKAN its dynamics can be approximated by those of VN and OKAN slow phase velocity in darkness. The saturation velocity and the time constant of OKAN are identical to those of the storage mechanism. The storage mechanism is also activated during OVAR.

In the Robinson model, storage is realized by an internal positive feedback loop created by an efferent copy of an eye velocity command signal. This loop has a time constant similar to that of the semicircular canals and its function is to complement the bandwidth of the canals during head rotation. It is unclear which neurons are responsible for this feedback loop and whether these neurons belong to just one cell population or not. In support of the concept of two dynamically different "pathways" are the findings of single cell recordings and of lesion studies which will be reported below. Not clear at present is the relationship between the smooth pursuit system and the "direct" pathways of OKN, and what the characteristics of visual information are that activate these systems. SP is induced by foveal and extrafoveal stimulation. However, SP does not activate the "indirect" pathways in monkeys as shown by the lack of any after-response. In monkeys with even very large central scotomata the "direct" pathways, and probably also the SP system, are still activated (Büttner et al. 1983). Nevertheless, central stimuli are more powerful than peripheral stimuli to activate the "direct" pathways.

SP eye movements are usually made in light. Tracking of a target in front of a stationary background could therefore rapidly activate the "direct" pathways. OKN would then be induced with the direction of slow phases opposite to the SP movements as soon as the target is followed. Humans and monkeys, however, can easily perform SP in light (Collewijn et al. 1982). This suggests that foveal motion information can override motion information from the peripheral retina. The underlying neural basis for the decision, volitional or not, to foveate a single moving target instead of stabilizing gaze according to peripheral retinal input is unknown.

The second important aspect of models on visual-vestibular interaction is how visual suppression of vestibular nystagmus and optokinetic after-nystagmus is realized. When a moving target is pursued with the head or when the animal is rotated passively together with an animal-stationary visual pattern the eye velocity command signal of the VOR has to be cancelled. It is generally held that this cancellation is effected, although not exclusively, by the smooth pursuit system (Robinson 1981b). It is also assumed that the cancelling signal from the smooth pursuit system and the VOR signal combine not at the level of the second-order vestibular neurons but at the level of the oculomotor neurons (Robinson 1981a, b). The question arises whether visual fixation does influence the characteristics of the velocity storage mechanism during VN and OKAN. Raphan et al. (1977, 1979) could demonstrate with their paradigm of giving short periods of full-field fixation during VN and OKAN (see Sect. 2.6.2) that indeed fixation not only rapidly cancels eye velocity but that it also rapidly reduces or "dumps" activity within the storage mechanism. Recently we showed that fixation of a single target light has similar effects on the storage mechanism as full field fixation during VN and OKAN (Waespe and Schwarz 1986). In the model of Raphan and Cohen the effects of visual fixation on the "eye velocity storage" integrator is incorporated by a switch mechanism which is activated by full field and also by single target fixation. In Robinson's model the positive eye velocity feedback loop is attenuated during fixation as eye velocity is strongly diminished. As efference feedback is blocked during the period of fixation activity of the storage mechanism is less after fixation than if there were no fixation period. However, in Robinson's model any eye velocity, regardless of how it is elicited, charges the storage mechanism. This is not supported experimentally as SP does not lead to a significant after-response in monkeys (Lisberger et al. 1981). It is unknown how the central nervous system actually realizes suppression of nystagmus or "dumping" of activity responsible for VN and OKAN. The nodulus and uvula seem to play a crucial role (Waespe et al. 1985a; Sect. 6.3.1).

2.8 Comment

The interaction of visual and vestibular information for gaze stabilization, although its analysis in this review is restricted to the horizontal plane only, is rather complex. The complexity will be even greater if gaze stabilization is analyzed in all three dimensions of space. In addition, dynamics of vertical and torsional eye movements are different from horizontal movements. In order to arrive at a full understanding of vestibular- and visual-oculomotor transformations, results of visual-vestibular interaction for gaze stabilization in the horizontal plane cannot necessarily be generalized to the second and third dimensions. Research in this direction is underway, but it imposes entirely new problems in terms of experimental apparatus, eye movement measurements, and complexity of analysis. As a first step we mentioned in Sects. 2.2.1 and 2.2.2 how otoliths and vertical canal stimulation lead to or modify horizontal nystagmus. This suggests that neuronal organization is not completely separate for the three dimensions. We have restricted our discussion to slow eye movements, but fast eye movements also contribute to gaze stabilization. It is unknown how fast eye movements, either during nystagmus where they are anticompensatory, or during SP where they are compensatory, are triggered.

3 Peripheral Vestibular System

The sense of hearing and the sense of balance have similar types of receptor cells, the hair cell. Hair cells of the vestibular end organ respond to extremely small displacements of the hair bundle (Hudspeth 1983). The mechanoelectric transduction process, i.e., the conversion of a mechanical force into an electric signal, and the microanatomy of the vestibular end organ have recently been reviewed (Wilson and Melvill Jones 1979).

The semicircular canal planes in monkeys have been measured with respect to the stereotaxic coordinate system (Blanks et al. 1985; Reisine et al. 1985). Canal planes on one side are roughly orthogonal to each other, and synergistic pairs of canal planes are approximately parallel with angles of $2^{\circ}-12^{\circ}$ between pairs in the rhesus monkey and $13^{\circ}-16^{\circ}$ in the squirrel monkey. The lateral canals form an angle of $15^{\circ}-22^{\circ}$ with the horizontal stereotaxic plane in the rhesus monkey and 18° in the squirrel monkey. A head position of 20° pitch nose-down would therefore produce maximal activation of the lateral canals and at the same time little activation of vertical canals during horizontal angular acceleration about a space-vertical axis.

Primary afferent fibers in the eighth nerve contact chemically hair cells of the ampullae of the three semicircular canals and of the otoliths and project to the vestibular nuclei in the brainstem. In the cat all vestibular nuclei except the dorsal part of the lateral nucleus receive inputs from primary vestibular neurons (Korte 1979). Fibers also project to parts of the cerebellum (A. Brodal 1981; monkey, Carleton and Carpenter 1984; cat, Korte and Mugnaini 1979) and also to the reticular formation (Korte 1979).

The average resting discharge recorded as fiber activity in the eighth nerve is about 100 $imp \cdot s^{-1}$ (range 18–180 $imp \cdot s^{-1}$) in the alert as well as in the anesthetized monkey (Goldberg and Fernandez 1971a; Keller 1976; Büttner and Waespe 1981; Louie and Kimm 1976; Miles and Braitman 1980). Robinson (1981b) has pointed out that a change in the resting discharge of only 2% in one canal would induce spontaneous nystagmus of about $2^{\circ} \cdot s^{-1}$. According to the regularity of firing, up to three fiber types can be distinguished. Regularly firing fibers have a coefficient of variation (standard deviation/mean spike interval) below 0.1 and irregularly firing fibers above, up to 0.8 (Goldberg and Fernandez 1971a; Keller 1976; Büttner and Waespe 1981; Louie and Kimm 1976). Regularly firing fibers tend to be of small diameter with a lower propagation velocity. Activity is not modulated with fast eye movements or different eye positions (Keller 1976; Büttner and Waespe 1981).

3.1 Vestibular Stimulation

All fibers connected to the horizontal canals are activated with angular acceleration to the ipsilateral side and inhibited with acceleration in the opposite direction (Fig. 13A). According to the nomenclature of Duensing and Schaefer (1958) this corresponds to a type I behavior. The sensitivity to acceleration is on average 0.4 imp·s^{-1} per $^{\circ}$·s^{-2} at 0.5 Hz in the alert monkey (Keller 1976) and is similar to that in the anesthetized monkey (Goldberg and Fernandez 1971a, b). The irregular neurons are more sensitive than regular neurons. With accelerations above 250°·s^{-2} in the inhibitory direction, average firing is silenced and the neurons are driven beyond their linear working range. At 0.2 Hz (± 20°) the sensitivity to head velocity averaged 0.65 imp·s^{-1} per $^{\circ}$·s^{-1} and the phase lead was 19° for lateral canal neurons (Miles and Braitman 1980). Neuronal

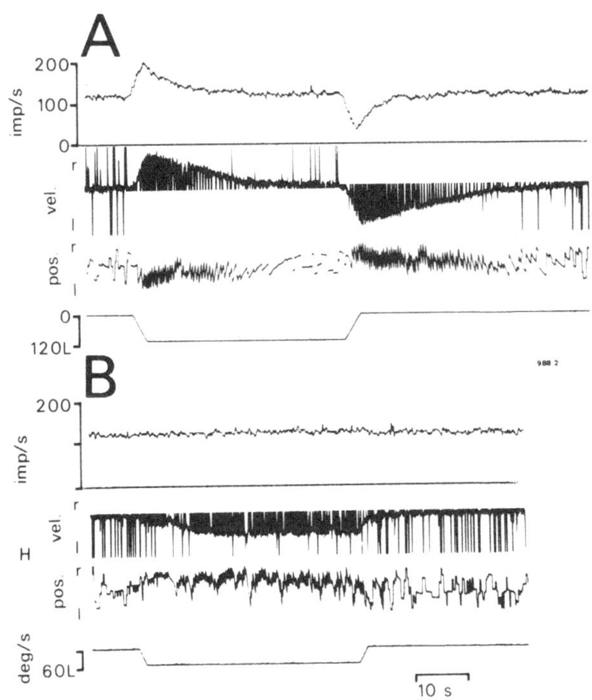

Fig. 13A, B. Firing rate of a vestibular fiber in the eighth nerve recorded in the alert monkey. *First line* in each subfigure, neuronal activity (running average of 250 ms; *second line,* velocity of horizontal eye movements; *third line,* horizontal eye position; *last line,* stimulus velocity profile (acceleration and deceleration 40°·s^{-2}, constant velocity 160°·s^{-1}. **A** During vestibular stimulation, i.e., rotation of the monkey in complete darkness; **B** during optokinetic stimulation (Büttner and Waespe 1981)

activity is in phase with head acceleration below 0.05 Hz (Fernandez and Goldberg 1971). Further details of adaptive effects and the behavior of primary fibers at frequencies of head rotation above 1.0 Hz are described in Goldberg and Fernandez (1975). Blanks et al. (1975) for the cat and Reisine et al. (1985) for the Rhesus monkey showed that activity is attenuated according to a cosine function when the horizontal canal is tilted away from its optimum plane of activation. In response to a sudden step in velocity, the latency of modulation ranges between 5 and 17 ms (Lisberger and Pavelko 1986; Miles et al. 1980b). After the end of an acceleration pulse, neuronal activity returns to the level of resting discharge with a time constant (Tc) of about 6 s (range 3–7 s) in the alert and in the anesthetized monkey (Fernandez and Goldberg 1971; Büttner and Waespe 1981). Thus the time constant of vestibular nystagmus is three to four times longer than that of primary vestibular neurons.

3.2 No Effect of Optokinetic Stimulation

Modulation of horizontal primary fibers to vestibular stimulation is independent of the presence or absence of concomitant visual stimuli Optokinetic stimulation with sinusoidally moving stimuli at 0.5 Hz (Keller 1976) or with constant stimulus velocities between 15 and $160^{\circ} \cdot s^{-1}$ (Büttner and Waespe 1981) did not modulate activity in vestibular nerve fibers (Fig. 13B; cat, Blanks and Precht 1978).

3.3 Off-vertical Axis Rotation

Constant velocity rotation about an axis tilted from the space-vertical (OVAR) produces a horizontal direction-specific nystagmus (see Sect. 2.2.2). During such constant velocity rotation there is no angular acceleration and the only vestibular stimulus is the rotating gravity vector which induces a sinusoidally modulated signal in otolith afferents (Fig. 14B). During OVAR neither semicircular nor otolith afferents give rise to a directionally specific constant increase or decrease in activity (Fig. 14; Raphan et al. 1983b; Goldberg and Fernandez 1982). Therefore it has been postulated that otolith information is further processed centrally to induce the direction-specific horizontal eye velocity signal (Raphan et al. 1983b). In agreement with this interpretation, horizontal nystagmus during OVAR is lost in the rabbit after the utricular nerves have been cut (Janeke et al. 1970).

In the alert monkey we found the horizontal canal afferents not to be sensitive to the static tilt of the animals (Raphan and Waespe, unpublished; Reisine, unpublished).

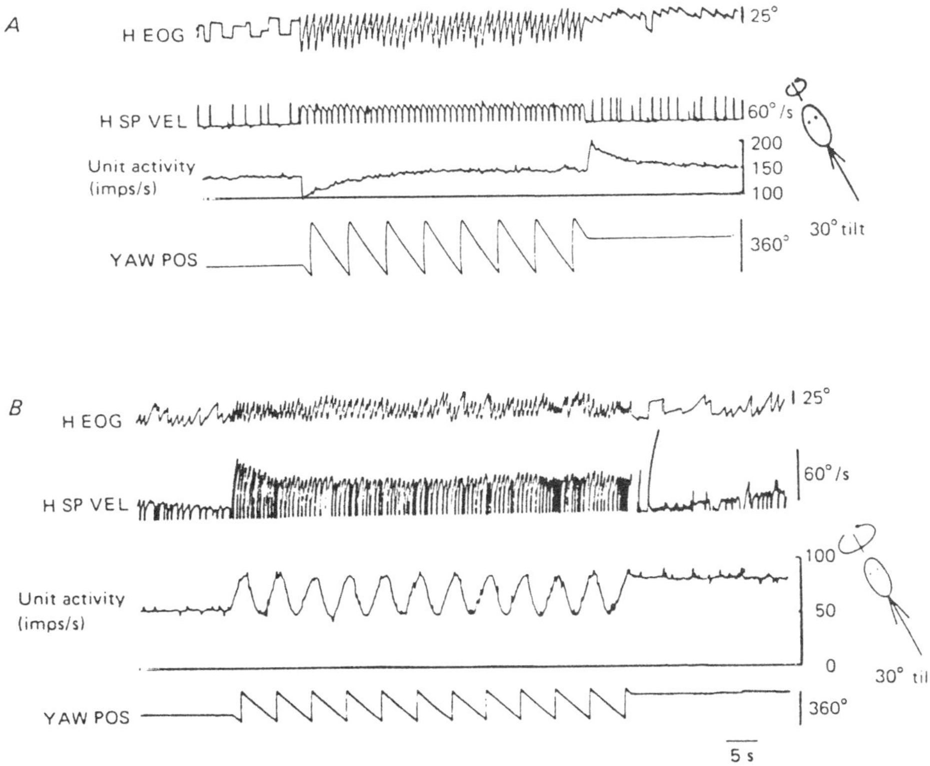

Fig. 14A, B. Responses of a regularly firing horizontal canal neuron (**A**) and a regularly firing otolith neuron (**B**) to a step in head velocity during off-vertical axis rotation (OVAR). The *traces* in **A** and **B** from *top* to *bottom* are: horizontal EOG, velocity of the horizontal eye movements, averaged firing rate of the neurons (counting interval 192 ms, updated every 1.6 ms) and angular position of the head. Note the increase in the firing rate of the canal neuron to its spontaneous level while nystagmus continues (**A**). The otolith neuron is continuously modulated with peaks in firing rate at a specific head position (**B**) (Raphan et al. 1983b)

3.4 Efferent Innervation

Like hair cells of other systems those of the labyrinths also receive an efferent innervation (Lindeman 1970; Gacek and Lyon 1974). In the squirrel monkey there are 400–600 efferent fibers in each vestibular nerve. They originate on both sides of the brainstem mainly from a region lateral to the abducens nucleus (Goldberg and Fernandez 1980). With 18 000 afferent fibers (Gacek and Rasmussen 1961) the ratio of efferent to afferent fibers is about 1:36. In the anesthetized squirrel monkey Goldberg and Fernandez (1980) found a predominantly excitatory action on afferent fibers with high frequency stimulation of efferent pathways. Together with natural stimulation there was mostly a decrease of sensitivity, especially in irregular neurons. They proposed that the efferent system could control activity of the vestibular receptors during active head movements.

3.5 Labyrinthine Lesions

Techniques to eliminate the information originating in the vestibular end organ include surgical destruction of the labyrinths (labyrinthectomy); section of single nerve branches peripheral to Scarpa's ganglion (neurectomy); section of the vestibular part of the eighth nerve proximal to the ganglion (neuronectomy); or plugging the semicircular canals by obstructing the endolymph flow (canal plugging). With all different lesion techniques the vestibular response originating in and specific to the lesioned part is abolished with some important differences between the effects of labyrinthectomy or neur(on)ectomy versus canal plugging. After canal plugging the crista and ampulla appear normal histologically and the tonic activity in the afferents remains intact (Goldberg and Fernandez 1975). Labyrinthectomy or neur(on)ectomy abolish most or all of the tonic resting discharge activity of primary afferents. In the guinea pig with lesions distal to Scarpa's ganglion some spontaneous activity remained, albeit on a much reduced level (Jensen 1983; Sirkin et al. 1984). When the proximal nerve stump was stimulated, eye movements could still be elicited. Also, even 1 year after labyrinthectomy, up to 75% of the cell bodies in Scarpa's ganglion show morphologically no degenerative changes (Schuknecht 1982).

After plugging of synergistic canal pairs, only the response specific to the affected canals is lost. As there is no asymmetry in afferent nerve discharge at rest, there is no spontaneous nystagmus, even immediately after surgery. Nystagmus during OVAR is preserved as it is thought to originate from otoliths' afferent activity. OKAN is also preserved, although gain is sligthly reduced (Cohen et al. 1983). On the other hand, after labyrinthectomy or neur(on)ectomy an important modification of visually induced eye movements is observed. OKN is slowed down, more irregular; and primary as well as secondary OKAN are permanently abolished (Cohen et al. 1983, 1973; Waespe and Wolfensberger 1985). The slow rise in OKN can no longer be evoked. Monkeys still reach peak OKN velocities of up to $100-130^{\circ}\cdot s^{-1}$ (Fig. 15). As primary vestibular neurons do not code visual information, the effects of neur(on)ectomy or labyrinthectomy on OKN and OKAN must be indirect. It is likely that the secondary neurons in the vestibular nuclei need the continuous tonic input activity of the afferents in order to function

Fig. 15. Peak slow phase velocity of primate OKN during constant velocity rotation before and 12 days after bilateral neurectomy. The *open* and *filled circles* are averages of the five fastest slow phases of OKN; SDs are indicated for values at 60 and $120^{o}\cdot s^{-1}$ stimulus velocity

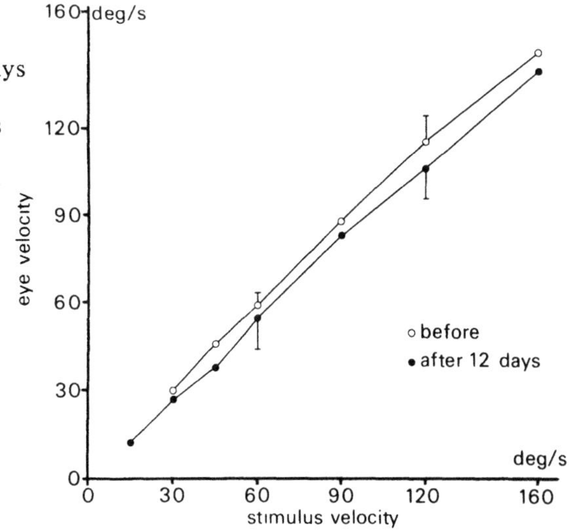

properly and to maintain their own resting discharge. This suggests that the peripheral vestibular apparatus determines how activity originating in the visual system is processed centrally even if it does not itself code visual information or visually induced eye movement information.

Recovery in eye-head coordination following bilateral labyrinthectomy in the monkey has been examined by Dichgans et al. (1973a). At least three strategies have been found to stabilize gaze: enhancement of the gain of the cervico-ocular reflex, central preprogramming of slow eye movements, and recalibration of saccades in response to retinal error signals (for humans see Kasai and Zee 1978).

3.6 Summary

The resting discharge of primary vestibular afferents is similar in alert and anesthetized monkeys. Primary afferents are not influenced by visual stimulation and their modulation in response to vestibular stimulation is the same regardless of the presence or absence of visual stimuli. Time constants of afferents are three to five times shorter than those of vestibular nystagmus. To account for the nystagmus response during visual-vestibular interaction and during OVAR, substantial central information processing is necessary. The role of the efferent vestibular system in alert monkeys remains enigmatic. It seems to play no part during visual-vestibular interaction, either on a short-term or long-term basis.

4 Visual Motion Information

Visual information to elicit slow eye movements can reach the brainstem via cortical projections, or directly via subcortical pathways. A direct retinal projection to the pretectum and accessory optic system (AOS) has been shown to exist in all mammals examined (Scalia 1972). This projection seems to be important for the generation of OKN as lesions in the pretectum abolish, and electrical stimulation induces, OKN (rat, Precht et al. 1982; rabbit, Collewijn 1975; monkey, Kato et al. 1986).

4.1 The Pretectal Complex and the Accessory Optic System

The pretectal complex with the olivary nucleus and the anterior, medial, and posterior pretectal nuclei (Hutchins and Weber 1985; Weber 1985; Scalia 1972; Lin and Giolli 1979), the accessory optic system (AOS) with the dorsal (DTN), medial (MTN), and lateral (LTN) terminal nuclei, and also the nucleus of the optic tract (NOT; Simpson 1984, Hutchins and Weber 1985) receive direct projections from the retina. All these nuclei except MTN lie close together near or within the brachium of the superior colliculus. It is not yet clear whether the NOT and the DTN form a continuous nucleus as suggested by physiological studies (Grasse and Cynader 1984). There are some interspecies differences in the retinofugal projection to the pretectum (Weber 1985). It has been questioned whether the monkey has a "complete" AOS (Simpson 1984; Weber 1985; Weber and Giolli 1986). In the cat and monkey, NOT and the pretectal olivary nucleus receive a bilateral projection from the retina and a projection from the cortical visual areas. In contrast, almost all projection to the AOS is contralateral (Lin and Giolli 1979; Giolli 1963). Physiological studies suggest that the AOS also receives an input from the visual cortical areas. In the cat, Marcotte and Updyke (1982) found projections from the suprasylvian visual areas to the terminal nuclei of the AOS. The "visual" pretectal regions project to the inferior olive, the nucleus reticularis tegmenti pontis (NRTP), and to the perihypoglossal region (Precht et al. 1982). There seems to be no direct projection to the cerebellum and to the pontine nuclei (Benevento et al. 1977; Berman 1977). In the monkey projections of the pretectum are largely unknown.

The physiology of the pretectal complex is not understood; a fair amount of data has been accumulated about the functional role of the AOS and NOT in the rabbit and cat, but little in the monkey. In the following we will discuss the properties of the DTN and NOT neurons in the cat.

In the cat neurons are driven best by moving large textured stimuli (Grasse and Cynader 1984). Neurons in the DTN or LTN are direction selective and respond to movements in the horizontal or vertical plane, respectively. Their receptive fields are large, with an average diameter of 60°, and always include the area centralis (Grasse and Cynader 1984). The inhibitory and excitatory vectors are separated by approximately 180° in the cat, but not in the lateral-eyed rabbit where they are noncollinear (Simpson 1984). Neurons in the DTN have.broad tuning curves between 1 and $100^{o} \cdot s^{-1}$ with strongest excitation in the velocity range of $6–15^{o} \cdot s^{-1}$. Almost all neurons are driven through the ipsilateral and contralateral eye although the retina projects mainly to the contralateral DTN. After ablation of the visual cortex, the DTN (and LTN) neurons became almost completely dominated by the input from the contralateral eye. Furthermore, the response to high stimulus velocities was abolished (Grasse and Cynader 1984). This strongly suggests that neurons in the DTN (and LTN) receive an input from the retina and from the visual cortex, in a manner similar to neurons in the NOT. The input from the visual cortex is obviously decisive for binocular responses and the modulation to high stimulus velocities of DTN neurons. In the rabbit the AOS receives no input from the visual cortex.

After ablation of the visual cortex in the cat, the monocularly induced OKN is much reduced for nasotemporal, but not for temporonasal, stimulation. The OKN asymmetry is then comparable to that in the normal rabbit. On the other hand, after chiasma section cats show a very weak OKN for patterns moving temporonasally, whereas for patterns moving nasotemporally OKN was only slighty affected (Harris et al. 1980). Thus, in the cat monocularly induced OKN with stimulation in the nasotemporal direction seems to depend on ipsilateral, uncrossed visual cortical pathways, and OKN with stimulation in the temporonasal direction seems to depend almost exclusively on crossed retinal pathways which directly connect to the pretectal region.

Neurons of the nucleus of the optic tract (NOT) which lie near the DTN also respond to large moving visual stimuli (cat, Hoffmann and Schoppmann 1975; rabbit, Collewijn 1975; monkey, Hoffmann and Distler 1986). Neurons in the DTN and NOT have very similar if not identical physiological properties (Simpson 1984; Grasse and Cynader 1984). In the cat the NOT neurons are also driven binocularly. After ablation of the visual cortex this drive is reduced to the contralateral eye only (Hoffmann 1982).

Hoffmann and Distler (1986) recorded the activity of NOT neurons in the anesthetized squirrel monkey. The neurons could be activated by stimulation of both eyes. The receptive fields were large with diameters up to 40° and always included the central region with the fovea. Neurons were excited with ipsilaterally moving visual stimuli and responded best in the velocity range of $10-80^{o} \cdot s^{-1}$. Most important and different to the cat, the neurons were equally well modulated by a movement of a small light spot of less than 1° if it was moved across the fovea.

Electrical stimulation in the region of the NOT induces OKN which is followed by an after-response comparable to OKAN in the rabbit (Collewijn 1975), cat (Hoffmann 1982), rat (Precht et al. 1982), and monkey (Cohen, personal communication). These eye movements are indistinguishable from naturally induced OKN and OKAN. Lesions of this region in primates abolish horizontal OKN with slow phases to the lesion side, leaving the OKN with slow phases to the opposite side and vertical OKN intact (Kato et al. 1986). Smooth pursuit eye movements were normal. Unfortunately the monkeys used in this study never showed an initial fast jump of OKN preoperatively, and OKAN saturation velocities were also not measured. The stimulus used by the authors (projected vertical stripes) obviously activated only the "indirect" visual-oculomotor pathways. These experiments suggest that the NOT lies in the "indirect" pathways but probably not in the "direct" oculomotor pathways as the smooth pursuit was intact. However, this is not yet clear because NOT neurons in the monkey respond to movements of large visual fields as well as of small targets. The latter does not activate the "indirect" pathways as there is no after-response after smooth pursuit (see Sects. 2.3.4, 2.7).

4.2 The Nucleus Reticularis Tegmenti Pontis

The nuclei of the AOS and NOT project to the nucleus reticularis tegmenti pontis (NRTP) (Precht et al. 1982; Maekawa et al. 1984), and the NRTP projects to the flocculus (cat, Hoddevik 1978; rabbit, Maekawa et al. 1981a, b). It is not yet clear whether the NRTP also projects to the vestibular nuclei (rabbit, Balaban 1983) or not (rat, Cazin et al. 1982). In the cat Gerrits et al. (1984) reported that the NRTP does not project to the lower brainstem. After NRTP lesions in the cat, OKN was severely impaired and eye velocity saturated at $10^{o} \cdot s^{-1}$ (Kato et al. 1982). Neurons of the NRTP or in its neighborhood responded to a moving visual pattern (rabbit, Maekawa et al. 1984; rat, Precht et al. 1982; monkey, Keller and Crandall 1983). In the monkey neurons were strongly activated at surround velocities of $5-10^{o}\cdot s^{-1}$ and saturated for higher velocities but still were modulated up to $100^{o}\cdot s^{-1}$. During smooth pursuit 30% of the neurons were also activated, suggesting that they responded to retinal slip and to eye velocity. However, they were not modulated during OKAN when retinal slip was absent. Some neurons were also modulated by vestibular stimulation. A large percentage of neurons has long-lead burst characteristics related to rapid eye movements.

4.3 Pontine Nuclei

Large parts of the visual cortex project to pontine nuclei (monkey, P. Brodal 1978; Glickstein et al. 1980; cat, Glickstein et al. 1972). In the monkey there is a dense projection traced by degeneration staining and HRP methods from the extrastriate cortex (both banks of the superior temporal sulcus, both banks of the intraparietal sulcus, and the anterior bank of the parieto-occipital sulcus) to the rostral, dorsolateral area of the pontine nuclei (Glickstein et al. 1980). The corticopontine projection appears to terminate in different regions of the pontine nuclei in monkey and cat (Glickstein et al. 1980). Neurons in the pontine nuclei of the cat are sensitive to moving visual stimuli (Baker et al. 1976). In the monkey the projection is to the dorsolateral peduncular, dorsolateral, and lateral nucleus. Projections from the superior colliculus seem not to overlap those from the cortex. In the cat there is also a dense projection from the visual and visual-associative cortex to the pars basalis of the pons (Albus et al. 1981).

In the alert monkey, Suzuki and Keller (1984) recorded activity in the dorsolateral pontine nucleus (DLPN) during smooth pursuit eye movements. Both, in the presence or absence of eye movements many cells responded to the velocity and direction of a random dot background pattern. Some neurons responded also to small moving targets of $1-2^{o}$ diameter.

4.4 Conclusions

"Visual" pathways to vestibular nuclei and the vestibulocerebellum via pretectal structures and pontine nuclei have been traced anatomically and electrophysiologically mainly in lower species (Glickstein and May 1982). It is certain that in the monkey there are also several regions in the pretectum and pons receiving a direct retinal input, which often overlaps with input from the visual and visual-associative cortex. The physiological properties of these structures in the pretectum and pons are largely unkown in the monkey. Furthermore, it remains unclear how they are connected to the vestibular nuclei and the flocculus which are known to mediate visually induced, slow eye movements command signals (see Sects. 5 and 6).

5 Vestibular Nuclei

Mainly four nuclei, the superior, medial, lateral, and inferior, constitute the vestibular nuclear complex. Cells within these nuclei are contacted by afferent fibers from the labyrinths and are called secondary vestibular neurons. Termination sites of afferent fibers are distributed unequally. Vertical canal fibers preferentially project to the superior nucleus, lateral canal fibers to the medial nucleus and ventral part of the lateral nucleus, and otolith fibers to the descending nucleus and the y-group (Walberg et al. 1958; Wilson et al. 1966), but their terminals are more scattered than previously thought (Korte 1979). In addition, primary vestibular fibers in the monkey were traced to the reticular formation, the vestibulo-cerebellum (see Sect. 6.1), the ascending cuneate nucleus, and the sub-trigeminal lateral reticular nucleus (Carleton and Carpenter 1984), but not to the prepositus nucleus.

The main projection site from the vestibular nerve to the superior vestibular nucleus (SVN) is its central core (A. Brodal 1984; Abend 1977; cat, Korte and Friedrich 1979). In turn it is this part which mainly projects to the oculomotor nucleus (Steiger and Büttner-Ennever 1979). The same central area is said to be the main projecting and receiving site of the reciprocal connection with the flocculus (monkey, Langer et al. 1985a, b; cat, Sato et al. 1983a, b). Mitsacos et al. (1983a, b), combining anatomical and electrophysiological methods, have found in the cat that neurons in the SVN receiving a monosynaptic excitatory input from the ipsilateral eighth nerve either project directly to the oculomotor nuclei or to the cerebellum, but not to both structures at the same time. The former cells generally do not receive a commissural inhibition and are located centrally, the latter cells are characterized by commissural inhibition and are localized more peripherally in the SVN.

The rostral part of the medial vestibular nucleus (MVN) is the main target site of the flocculus and projects mainly to oculomotor structures (Brodal 1984). Lateral to it is the main area where the vestibular nerve projection is particularly dense. The caudal part projects to the flocculus and receives cerebellar input from the nodulus in addition to spinal input.

The lateral vestibular nucleus (LVN) constitutes primarily the origin of the vestibulospinal fibers. It is not connected either efferently or afferently to the flocculus (monkey, Langer et al. 1985b; cat, Sato et al. 1983a). It will not be further considered in our context.

The inferior vestibular nucleus (IVN) shows a complex anatomical organization with various reciprocal cerebellar connections but little projection to the oculomotor system. Of the small cell groups the y-group projects heavily to oculomotor structures and in turn receives saccular input (Gacek 1969), thus constituting a direct otolith-oculomotor pathway.

This short anatomical survey shows that the rostral part of the medial nucleus and the central core of the superior vestibular nucleus contain cells directly projecting to the oculomotor system. Many but not all neurons, receiving a monosynaptic canal input, are reciprocally interconnected via a commissural system. Type I cells excite type II cells on the other side, which in turn inhibit type I neurons on the same side (Shimazu and Precht 1966; Precht 1978; Mitsacos et al. 1983a, b). For details of afferent and efferent projections of the vestibular nuclear complex in cat and monkey we refer the reader to Carleton and Carpenter (1983).

5.1 Vestibular Nuclei Activity

Single neuron recordings in alert monkeys pose practical restrictions concerning how far neurons can be identified. Neurons could be classified according to their monosynaptic input from primary vestibular units (Keller and Kamath 1975) or according to their efferent connections to the oculomotor nuclei (Tomlinson and Robinson 1984). However, this classification does not tell us much about the strength of the connections. Problems of anatomically defining subgroups of cells that constitute a nucleus have been discussed by Brodal (1972) and in recent literature reviewed by Brodal (1984) and Carleton and Carpenter (1983, 1984). A. Brodal (1981) especially discussed the problem that neurons often have a wide arborization of dendrites, extending beyond the anatomically defined boundaries of a vestibular nucleus. Only a few studies attempted to quantitatively estimate physiological cell types in the primate vestibular nuclei complex (Fuchs and Kimm 1975; Lisberger and Miles 1980). From these studies it is apparent that there are at least four to five different cell types which can be separated according to physiological criteria (for the vertical system see Tomlinson and Robinson 1984; Chubb et al. 1984). Other studies, including our own, have been biased in the search for cells which we have called "central vestibular neurons". These cells were recorded within the boundaries of the vestibular nuclei, were activated by head accelerations of 2.5 or $5^{o} \cdot s^{-2}$, and exhibited no or only a weak eye position sensitivity (below 1 $imp \cdot s^{-1}$ per ^{o}; Waespe and Henn 1977a, b, 1979, 1985).

Relating activity changes of vestibular nuclei neurons to the activation of semicircular canals can be complex. Head position is critical to determine whether a neuron receives horizontal canal input. Exact canal orientation has been discussed in Sects. 2.2.1 and 3. In the literature, head position is not always specified or is given as stereotaxic horizontal. During vertical axis rotation with the head in the stereotaxic horizontal plane, vertical canal neurons are consistently modulated. With vertical axis rotation alone one can not then differentiate between vertical canal or low sensitivity horizontal canal units. Type II "horizontal" units could well be type I vertical neurons.

During recent years it became apparent that a classification of neurons in the vestibular complex according to physiological criteria poses problems. The current classification is unsatisfactory as laboratories have usually tested the behavior of neurons only for selected experimental paradigms and not for all possible combinations of vestibular and visual stimulations. Also problematic is the classification according to signals as "head" or "eye" velocity, as will be discussed in further detail in Sects. 5.2.5 and 5.2.6. With these restrictions in mind we will describe the behavior of "central vestibular neurons" during visual-vestibular interaction.

5.2 Neurons Responding to Horizontal Canal Activation

The resting discharge varies between 1–160 $imp \cdot s^{-1}$ (Waespe and Henn 1979; Buettner et al. 1978). On average it is about 50% of the resting discharge found in primary vestibular neurons. The activity of about half of the neurons is modulated in the absence of any stimulus in relation to fast eye movements (Miles 1974; Fuchs and Kimm 1975; Keller and Daniels 1975; Keller and Kamath 1975; Waespe et al. 1977). Accordingly, neurons have been grouped into vestibular-plus-eye-movement and vestibular-only neurons. This terminology is unsatisfactory as all "vestibular-only neurons" are also modulated during OKN and OKAN (Sects. 5.2.2 and 5.2.3). Some vestibular-plus-eye-movement neurons pause with saccades in all directions (omnipause cells) or only in a specific direction. Others show bursts in one and pauses with saccades in the opposite direction. Other vestibular-plus-eye-movement units show in addition some modulation with different eye positions. The usual pattern is an increase in activity with fast movements and positions in the direction opposite to that which excites the neuron during vestibular stimulation (Miles 1974). Central vestibular neurons have, if at all, only a weak eye position sensitivity below 1.0 imp$\cdot$s^{-1} per $^{\circ}$. Other neurons have a prominent eye position signal. They are mostly located in the rostral area of the medial vestibular nucleus (Keller and Kamath 1975; Keller and Daniels 1975).

5.2.1 Vestibular Stimulation

In response to angular acceleration about the earth-vertical axis in darkness, about two-thirds of neurons receiving a horizontal canal input show a type I response (nomenclature, Duensing and Schaefer 1958). They have a bidirectional modulation with an activity increase during angular acceleration to the ipsilateral side and a decrease to the contralateral side of recording. Many neurons are completely silenced, showing a cut-off during acceleration to the contralateral side due to the low resting discharge (Figs. 16, 19). One-third of the neurons show a type II

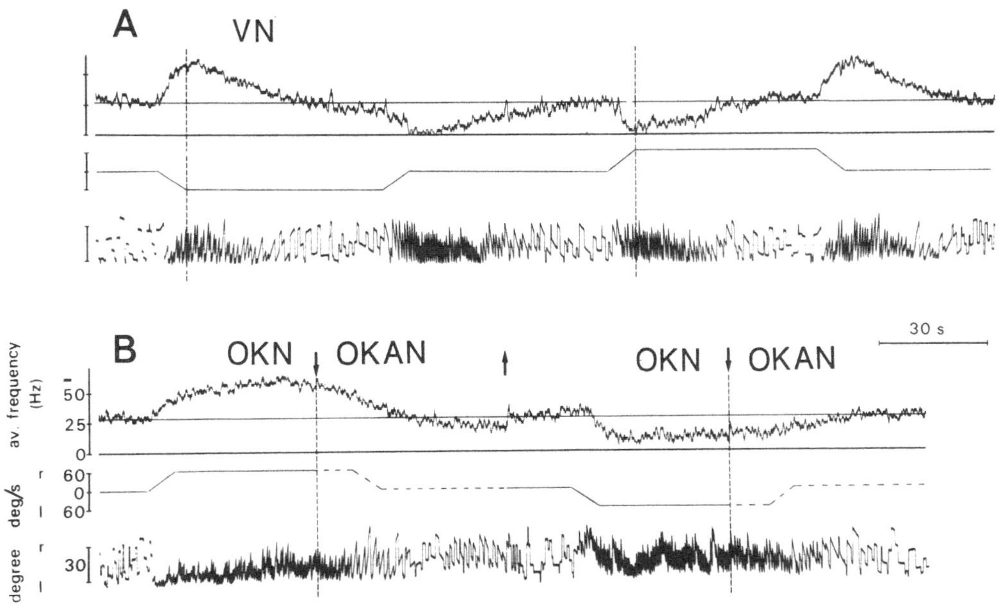

Fig. 16A, B. Firing rate of a type II neuron in the vestibular nuclei during vestibular (**A**) and optokinetic stimulation (**B**). *First trace,* neuronal activity (running average); *second trace,* stimulus profile (acceleration $8^{\circ}\cdot s^{-2}$, constant velocity $60^{\circ}\cdot s^{-1}$); *third trace,* horizontal eye position. *Downward arrow* marks the time when lights are switched off, *upward arrow,* lights on. The time course of neuronal activity changes and of eye velocity are similar during VN (**A**) and OKAN (**B**) (Waespe and Henn 1977b)

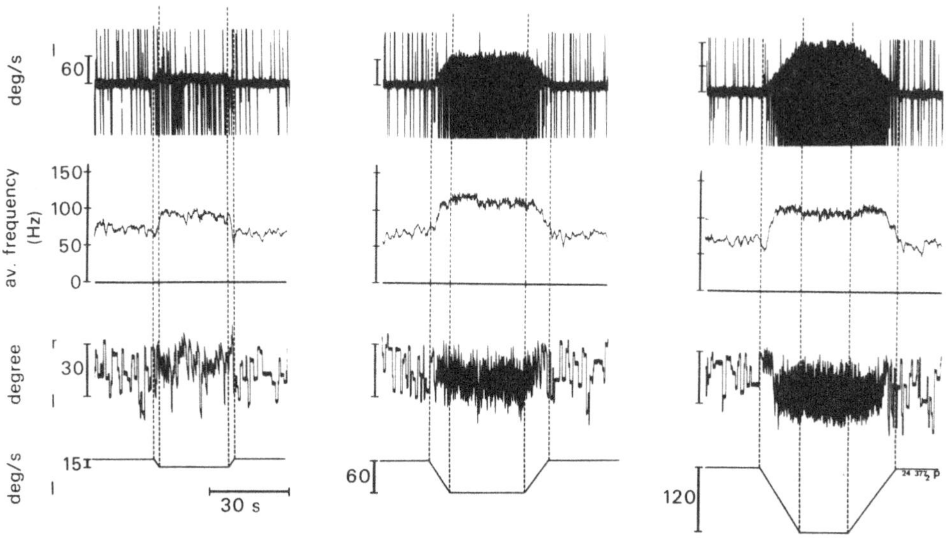

Fig. 17. Firing rate of a type I "central vestibular neuron" and nystagmus slow phase velocity during different optokinetic stimulus velocities of 15, 60, and $120^{\circ} \cdot s^{-1}$. Although nystagmus velocity increases from a stimulus velocity of 60 to $120^{\circ} \cdot s^{-1}$, neuronal activity does not. At $120^{\circ} \cdot s^{-1}$ stimulus velocity, frequency increases rapidly and reaches a steady state value already during the acceleration. This level of activity is also maintained for the initial 13 s of the deceleration until the velocities of the stimulus as well as of OKN fell below the saturation velocity of the neuron which was about $45^{\circ} \cdot s^{-1}$ (Waespe et al. 1977)

response with activity increase during acceleration to the contralateral side. The head of the monkey was tilted 20°–25° nose-down thereby minimizing the activation of vertical canals. Type I and II neurons behave similarly during all the different stimulus paradigms as described below save that they are activated by stimulation in opposite directions. Type III (frequency increase in response to acceleration to the ipsilateral and contralateral side), and type IV (frequency decrease in response to acceleration in both directions) neurons described in other species have not been documented in the monkey.

The sensitivity of vestibular neurons relative to stimulus velocity is on the average 0.75 $imp \cdot s^{-1}$ per $^{\circ} \cdot s^{-1}$ measured during sinusoidal stimulation (Lisberger and Miles 1980; Buettner et al. 1978), and comparable to 0.94 $imp \cdot s^{-1}$ per $^{\circ} \cdot s^{-1}$ measured in response to an acceleration pulse of $2.5^{\circ} \cdot s^{-2}$ (Waespe and Henn 1979). Type I vestibular-plus-eye-movement neurons are the most sensitive ones (Fuchs and Kimm 1975; Lisberger and Miles 1980; Waespe and Henn 1979). Type I vestibular-only neurons have a lower sensitivity (Fuchs and Kimm 1975; Waespe and Henn 1979). It is crucial to note that for all these measurements nystagmus was present and not suppressed although Lisberger and Miles (1980) use the term "head velocity" sensitivity. It is not excluded that neurons have also some "eye velocity" sensitivity. This is apparent at low accelerations and further discussed in Sect. 5.2.6.

In response to a velocity step (for practical matters, an acceleration exceeding $10^{o} \cdot s^{-2}$), peak neuronal modulation occurs at the end of acceleration. Afterwards, neuronal activity returns to baseline levels with a dominant time constant (T_c) of about 15 s which is similar to the time constant of vestibular nystagmus (Fig. 16). T_cs of both, vestibular nystagmus and neuronal activity decay have a range between 8 and over 50 s but always covary together. Thus, in experimentally naive animals with long nystagmus time constants, T_cs of central vestibular neurons also have similarly large values. Another way to determine time constants is to measure the phase and gain in response to sinusoidal stimulations (Buettner et al. 1978). Values determined with either method show a close correspondence of T_cs. Time constants (T_c) of vestibular nystagmus and of central vestibular neurons are about two to four times longer than values found for activity changes in the nerve. This shows that statements about variations of time constants and of frequency response curves are only meaningful if values for vestibular nystagmus have been measured simultaneously.

Buettner et al. (1978) also compared time constants of the activity of central vestibular neurons in alert and anesthetized monkeys. During anesthesia, the average time constants of central vestibular neurons fell from 24 and 10 s respectively in two monkeys to 6.7 and 4.7 s respectively. After recovery from anesthesia, values returned to their previous range.

Neuronal modulation shows adaptation during prolonged accelerattions. This is apparent for values below $10^{o} \cdot s^{-2}$ applied over several seconds (Figs. 18 and 19). Frequency increases approximately exponentially up to a steady state level. At an acceleration of $5^{o} \cdot s^{-2}$ the vestibular-plus-eye-movement type I neuron in Fig. 19 increased activity with a time constant (T_i) of 3 s during the acceleration. The vestibular-only type I neuron in Fig. 18 and recorded in the same monkey had an equivalent time constant (T_i) of about 14 s. For the two neurons the dominant time constant after the end of acceleration (T_c) was similar, between 15 and 17 s, and similar to T_c of VN. The time constant for activity increase (T_i) in 15 vestibular-only and vestibular-plus-eye-movement type I neurons during a $2.5^{o} \cdot s^{-2}$ acceleration step was on average 8.9 s (range 4.8–19). Time constants for nystagmus (T_i) measured simultaneously was 13.3 s (range 8.5–18 s). The decay time constant (T_c) for the same neurons was 10.3 s (range 6–19 s) and for VN 11.5 s

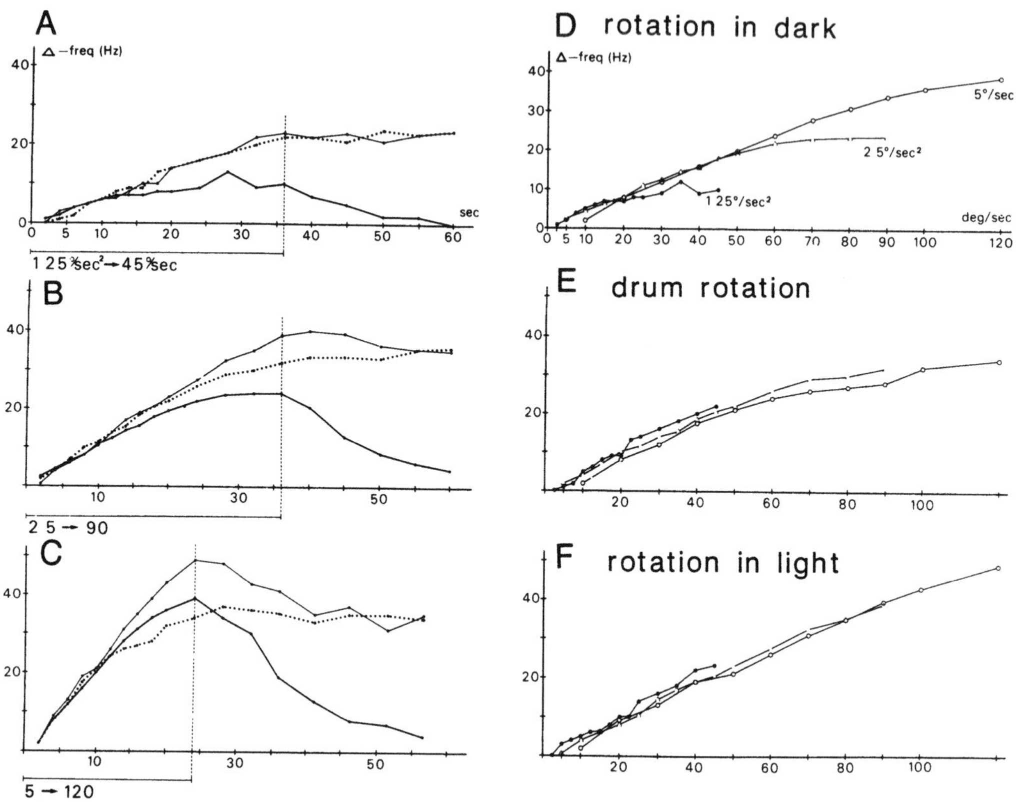

Fig. 18 A–F. Change in firing rate (*ordinate*) of a type I vestibular-only neuron in the vestibular nuclei against time (**A–C**) or instantaneous stimulus velocity during the acceleration phase (**D–F**). In **A–C** ***heavy line*** represents firing rate changes during vestibular, ***thin line*** during combined visual-vestibular, ***dotted line*** during optokinetic stimulation. ***Vertical dotted line*** indicates end of acceleration which is followed by constant velocity rotation. In **D–F** firing rate changes during accelerations are plotted against instantaneous velocity (in **D** for vestibular, in **E** for optokinetic, and in **F** for combined stimulation)

(range 8–20 s). The large values of T_i in vestibular-only neurons may reflect the activation of the velocity storage mechanism at these low accelerations.

Comparative studies in other species gave qualitatively similar results. Shimazu and Precht (1965) explored the brainstem of the cat, also finding about two-thirds of units to be type I, and one-third to be type II. As their and most other studies were performed in decerebrate or anesthetized animals (Shinoda and Yoshida 1974; Curthoys and Markham 1971; Melvill Jones and Milsum 1971; Schneider and Anderson 1976), this may explain the short time constants (T_c) of units in response to angular acceleration, and the absent modulation with eye movements. Sometimes it has therefore been stated that secondary vestibular neurons

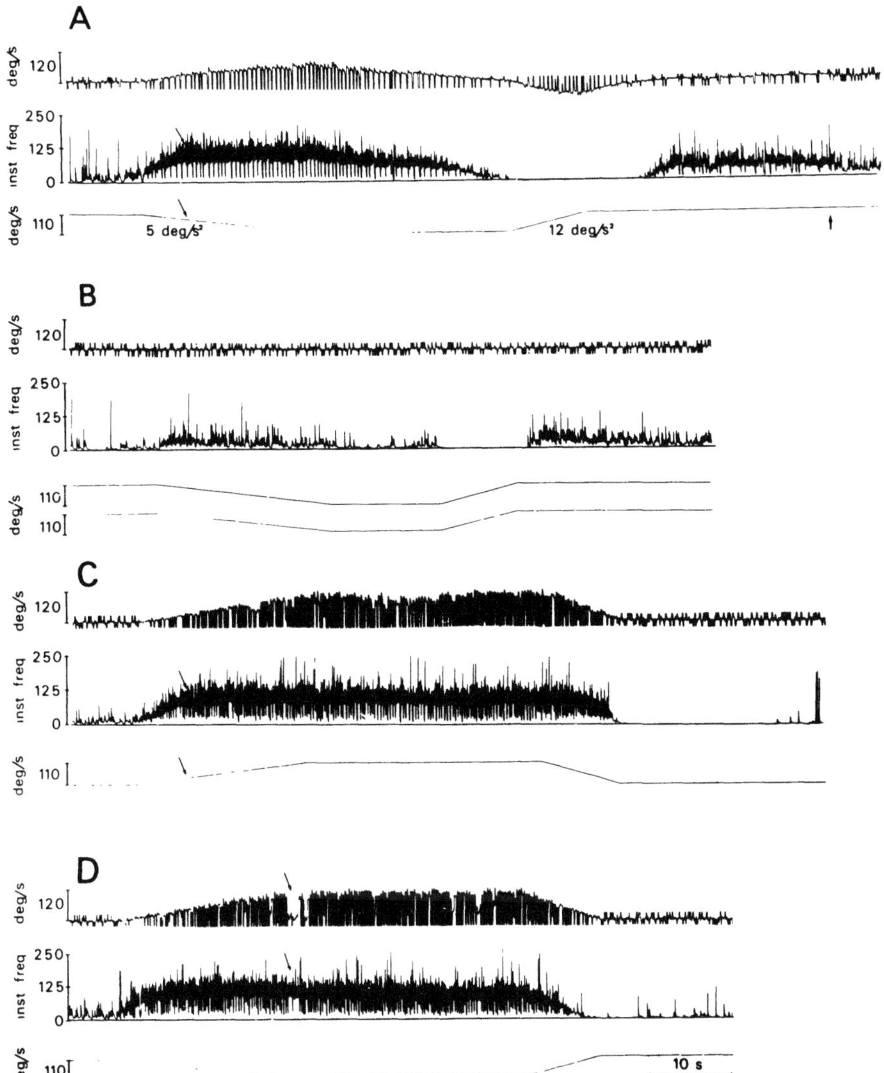

Fig. 19 A–D. Firing rate changes of a type I vestibular-plus-eye-movement neuron during vestibular (**A**), conflict (**B**), optokinetic (**C**), and combined (**D**) stimulation. *First trace,* eye velocity; *second trace,* instantaneous frequency; *third trace,* stimulus velocity (acceleration $5^{\circ}\cdot s^{-2}$, constant velocity $110^{\circ}\cdot s^{-1}$, deceleration $12^{\circ}\cdot s^{-2}$). When nystagmus decreases due to drowsiness of the monkey (**D**, arrow) neuronal activity does not change. However, when nystagmus is actively suppressed as during conflict stimulation (**B**) frequency modulation is minimal

carry essentially the same information as primary neurons in the vestibular nerve. From the work in alert monkeys there is now consensus that this view has to be modified, the two important aspects being the prolongation of the time constant and the frequent modulation of neurons in association with fast eye movements and eye positions. No quantitative analysis is available from the alert and behaving cat and rabbit, although the first experiments in alert animals were all done in the rabbit (Duensing and Schaefer 1958, 1960; Schaefer et al. 1977).

5.2.2 Optokinetic Stimulation

All central vestibular neurons are also modulated bidirectionally in response to full-field horizontal optokinetic stimulation (Figs. 16, 17, 19). Type I neurons increase activity with optokinetic stimulation to the contralateral side leading to OKN with fast phases to the ipsilateral side of recording. During optokinetic stimulation to the ipsilateral side the activity of type I neurons decreases. Type II neurons show a mirrorlike behavior (Fig. 16). Thus, whenever VN and OKN beat into the same direction, neuronal modulation either increases or decreases regardless of how nystagmus is induced. During optokinetic stimulation with constant velocities, modulation is maintained throughout stimulation (Figs. 16, 17). Neuronal modulation saturates at a stimulus velocity of $60^{\circ} \cdot s^{-1}$ on average (range $30–90^{\circ} \cdot s^{-1}$; Fig. 17), with the more sensitive neurons saturating at a lower level (Figs. 18, 19). In contrast, OKN slow-phase velocities saturate above stimulus velocities of $100^{\circ} \cdot s^{-1}$ (Figs. 8A, 10). During constant accelerations of the optokinetic stimulus below $5^{\circ} \cdot s^{-2}$ and end velocities not exceeding $60^{\circ} \cdot s^{-1}$ neuronal activity is on average proportional to the actual stimulus velocity. At higher accelerations and end velocities neuronal activity changes lag behind instantaneous stimulus or eye velocities (Waespe and Henn 1979). Consequently, it is expected that central vestibular neurons have a poor high frequency response to sinusoidal optokinetic stimulation. Boyle et al. (1985) stated that 87% of the neurons recorded in the vestibular nuclei did not respond to optokinetic stimulation above 0.1 Hz (max velocity $40^{\circ} \cdot s^{-1}$; max acceleration $50^{\circ} \cdot s^{-2}$) although OKN was still evoked. Only 15% of the neurons responded up to 1 Hz.

If during nystagmus eye velocity momentarily collapses because of drowsiness of the monkey, neuronal activity in many neurons stays at the same level (Fig. 19, arrow). Neuronal activity may therefore not be related to eye velocity per se but rather reflects the internal state of an eye velocity generator, propably of the velocity storage mechanism.

Comparative studies suggest that full-field optokinetic stimulation modifies activity in vestibular nucleus cells in all species tested exept in the frog. This has first been described in the goldfish (Dichgans et al. 1973b; Allum et al. 1976) and in the alert monkey (Henn et al. 1974). In paralyzed nonanesthized animals neuronal responses are tested under open-loop conditions and it is assumed that activity changes of vestibular nuclei neurons reflect image slip velocity, but other explanations cannot be ruled out. Given a state of muscle paralysis and with central structures not affected by the paralyzing drug, modulation could still reflect a slow eye velocity command signal which, however, leads to no OKN. An illustrative example is the behavior of vestibular nuclei neurons which are activated differently during optokinetic stimulation in the alert nonparalyzed and the alert paralyzed cat (Keller and Precht 1979). At low stimulus velocities neuronal modulation is about twice as high in the paralyzed cat (i.e., under open-loop conditions) as in the nonparalyzed behaving cat, and modulation increases monotonically up to stimulus velocities of $10^{\circ} \cdot s^{-1}$ only in the paralyzed cat instead of $40-60^{\circ} \cdot s^{-1}$ as in the nonparalyzed cat (i.e., under closed-loop conditions). In the rat (Precht et al. 1982) the response is unidirectional if the ipsilateral eye is covered. OKN or vestibular nucleus modulation can then only be evoked with temporonasal stimulation. With both eyes open, OKN became symmetric, and vestibular nuclei neurons responded to stimulation in both directions. This asymmetry has its reason in the anatomy of the visual pathways in the rat which has been worked out by Precht et al. (1982). So far the frog seems to be the only animal in which an optokinetic modulation of vestibular neurons could not be shown to exist (Dieringer and Precht 1982). These authors also discuss possible reasons in terms of comparative physiology and anatomy.

5.2.3 Optokinetic After-nystagmus

If during optokinetic stimulation the lights are switched off, nystagmus continues as OKAN. As described in Sect. 2.3.2, a fast initial drop in eye velocity is observed if the preceding OKN velocity exceeds the OKAN saturation velocity of about $60^{\circ} \cdot s^{-1}$. Activity of central vestibular neurons does not change rapidly but slowly reduces to its rest level. The time course of activity changes faithfully reflects the time course of OKAN velocity (Fig. 16; Waespe and Henn 1977b). The decay time constants (T_C) of OKAN and neuronal activity changes are comparable and similar to T_Cs for vestibular nystagmus and accompanying neuronal activity changes (Fig. 16). If the monkey is left long enough in darkness so that secondary OKAN develops, the change in nystagmus direction is also reflected in neuronal activity as an under- or overshoot (Fig. 16B; Waespe and Henn 1977b). If lights were turned on during primary OKAN and the monkey views the stationary visual surround, nystagmus is rapidly suppressed (Sect. 2.6.2). Neuronal activity then also returns to the resting discharge level within 2–4 s and primary OKAN does not develop after the lights are switched off again (Waespe and Henn 1977b).

The characteristics of unit activity during OKN with the slow build-up of activity and the slow discharge after lights off in parallel with OKAN velocity suggest that central vestibular neurons constitute a part of the "indirect" visual-oculomotor pathway (see Sect. 2.7). It also suggests that neuronal activity reflects an eye velocity command signal but not image slip velocity. As shown in Sect 5.2.5, this is not fully supported experimentally.

5.2.4 Rotation in Light

Animal rotation in the light most closely mimics natural conditions save that all movements are passive. As expected, neuronal responses for vestibular and optokinetic stimulation combine so that stimulus velocity is more faithfully reflected, but still within limits, in the activity changes of neurons (Robinson 1977). This is shown for a type I neuron in Fig. 18. Only when the animal is rotated in the light, a linear relation between unit activity and instantaneous stimulus or eye velocity during the acceleration phase is observed regardless of the duration or the value of the acceleration. During vestibular and optokinetic stimulation alone, the response strongly depends on the value and the duration of the acceleration. During rotation at a constant velocity, modulation is maintained at the same level as during optokinetic stimulation alone (Figs. 16, 18, 19). This shows that rotation in light leads to a response which combines the high frequency characteristics of the vestibular response and the low frequency characteristics of the optokinetic response to yield activity which reflects actual stimulus velocity if nystagmus is not suppressed (Waespe and Henn 1979).

5.2.5 Optokinetic Stimulation with Nystagmus Suppressed

Buettner and Büttner (1979) studied the behavior of central vestibular neurons during optokinetic stimulation while the trained animal fixated a small light target, thereby suppressing nystagmus. If suppression is complete, image slip velocity corresponds to stimulus velocity. Although this creates experimentally an open-loop situation, it most probably cannot be compared to stimulation under open-loop conditions in paralyzed animals (see Sect. 5.2.2). On average, neuronal activation is reduced during complete suppression of OKN to about 60% as compared to 100% during OKN at a stimulus velocity of $40^{\circ} \cdot s^{-1}$. This demonstrates that neuronal activity in the vestibular nuclei is not solely determined by nystagmus or image slip velocity.

Further evidence that activity changes of vestibular neurons does not simply reflect the slow phase velocity of nystagmus comes from data obtained from animals with experimentally induced unilateral gaze palsies. Jaeger et al. (1981) found optokinetic modulation in vestibular neurons still present, although the monkey with a unilateral lesion of the paramedian pontine reticular formation had a gaze palsy toward the ipsilateral side.

5.2.6 Visual-Vestibular Conflict-Stimulation and VOR-Suppression

In the paradigm of conflict-stimulation the monkey is rotated in the light while the optokinetic drum is mechanically fixed to the turntable. The accompanying nystagmus has been described in Sect. 2.6.2. At low accelerations when vestibular nystagmus is completely suppressed, activity changes of all central vestibular neurons is reduced compared to acceleration in darkness when there is VN (Fig. 19A, B). At higher accelerations above $10-20^{\circ} \cdot s^{-2}$, however, peak modulation for conflict-stimulation is similar to that for vestibular stimulation alone although nystagmus is reduced to less than 10% during conflict-stimulation. This experimental paradigm also demonstrates that the behavior of all central vestibular neurons is influenced by the occurrence of nystagmus, at least in the low acceleration range used in our studies. The threshold to vestibular stimulation is enhanced and the sensitivity to "head velocity" is lowered during conflict-stimulation compared to the values during vestibular stimulation alone. Thus, in order to define the signals conveyed by these neurons, the experimental conditions have always to be specified. The fact that most central vestibular neurons are not modulated at very low accelerations of 2.5 and $1.25^{\circ} \cdot s^{-2}$ during conflict-stimulation but strongly during vestibular stimulation alone suggests that these neurons mediate an "eye velocity" signal. However, at accelerations above $10^{\circ} \cdot s^{-2}$ these neurons are modulated as strongly during conflict-stimulation as during vestibular stimulation irrespective of strength of nystagmus, suggesting that they then convey a signal proportional to "head velocity". The time constants of neuronal decay are always below 6–8 s during conflict-stimulation, regardless of the strength of acceleration. The situation is even more complex as the observed attenuation of neuronal activation by conflict-stimulation is dependent on how nystagmus is suppressed. Buettner and Büttner (1979) have tested central vestibular neurons during conflict-stimulation, VOR-suppression, and vestibular stimulation. Attenuation in neuronal activation is only observed if nystagmus is suppressed by a full field but not by a single target light which the monkey fixates (VOR-suppression). The time constant of neuronal decay nevertheless was also shorter during VOR-suppression compared to vestibular stimulation but not as short as during conflict-stimulation.

5.2.7 *Smooth Pursuit*

Keller and Kamath (1975) and Keller and Daniels (1975) have tested cells in the vestibular nuclei during SP also. Neurons with no eye position sensitivity were found not to be modulated during SP. These neurons were excited monosynaptically by stimulation of the eighth nerve (average latency 0.86 ms). In contrast, cells with an eye position sensitivity were generally found to be modulated during SP in relation to eye velocity also. During VOR-suppression neuronal activity was not related to head velocity, suggesting that these cells do not mediate a "head" but only an "eye velocity" signal. These neurons were excited mostly disynaptically (average latency 1.36 ms).

A small region 0.5–1.5 mm below the nucleus abducens was described by Eckmiller and Mackeben (1980) to contain neurons which were modulated during SP in relation to eye velocity only. The activity of these same neurons was not modulated in relation to different eye positions.

5.3 Conclusions

Cells in the vestibular nuclei receiving their input from the horizontal canals and with no or a negligible eye position sensitivity have been investigated during various paradigms of visual-vestibular interaction. These cells have been called "central vestibular neurons" by us to differentiate them from cells with a strong eye position sensitivity. Results are consistent with the hypothesis that these "central vestibular neurons" mediate (besides a "head" velocity signal) signals of the "indirect" visual-oculomotor pathways and thus also reflect the activity of the eye velocity storage mechanism (see Sect. 2.7). Currently, it is unknown whether these "central vestibular neurons" project directly or only indirectly via burst-tonic neurons within the vestibular nuclei to the oculomotor neurons. Strong evidence exists that the "central vestibular neurons" project to the flocculus (Sect. 6.2.3.2). In recent years it became evident that the vestibular nuclei, taken as a single anatomical entity, subserve more than one function within the oculomotor system. Vestibular nuclei neurons do not only mediate "head" or "eye velocity" signals. Some reflect mainly the activity of the storage mechanism. There are cells in the rostral part of the vestibular nuclear complex which are characterized by a high sensitivity to eye positions. Their functions during OKN and OKAN are largely unknown. These neurons mediate an "eye velocity" signal during SP and VOR, but do not mediate a "head velocity" signal during VOR-suppression (Keller and Kamath 1975;

Keller and Daniels 1975). In addition, after chemical lesions in the anterior part of the medial vestibular nucleus and the perihypoglossal nuclei, the horizontal eye position integrator is severely disturbed if not abolished (Cannon and Robinson 1986). This suggests that neurons in the anterior part of the medial vestibular nucleus take part in the integration of eye velocity to eye position signals.

5.4 Nucleus Prepositus Hypoglossi

The nucleus prepositus hypoglossi (NPH), caudal to the nucleus abducens, has reciprocal connections to the vestibular nuclei and cerebellum. Cells also project to other oculomotor structures (McCrea et al. 1979; McCrea and Baker 1985a, b; A. Brodal 1983). Because of this anatomical location various speculations were put forward about its function in relation to vestibular and oculomotor signals (Baker and Berthoz 1975; Blanks et al. 1977). It has been mentioned as one of the relay stations for visual signals from the AOS to the vestibular nuclei; specifically, it receives monosynaptic projections from the superior colliculus and some of the pretectal nuclei (cat, Magnin et al. 1983). Speculations received further support from cell recordings in the paralyzed, alert rat in which vestibular as well as optokinetic responses could be observed (Lannou et al. 1984). This observation is supported by lesion experiments in which OKN as well as OKAN was lost or severely impaired (monkey, Uemura and Cohen 1973; cat, Cheron et al. 1986a, b). Single unit recordings in the alert cat revealed a majority of tonic or burst-tonic neurons related to eye position and parameters of fast and slow eye movements (Lopez-Barneo et al. 1982). Thus, cells in the NPH could be part of the "premotor" oculomotor organization, but their exact place in the causal chain of eye movement generation has not yet been determined.

In summary, all anatomical and electrophysiological evidence points towards the NPH as another structure involved in visual-vestibular interaction, but systematic single neuron recordings have not been done in the monkey.

6 Vestibulo-Cerebellum

6.1 Definition

Those parts of the cerebellum which receive a projection from primary vestibular neurons have been defined as the vestibulo-cerebellum (Brodal and Høivik 1964); they include the flocculus, ventral uvula, nodulus, and parts of the ventral paraflocculus. This definition has been questioned as being to restricted (Kotchabhakdi and Walberg 1978) or too generous (Korte and Mugnaini 1979; A. Brodal 1981). The lingula and folia V and VI of the vermis were also reported to receive a projection from primary vestibular fibers in the monkey (Carpenter et al. 1972).

6.2 Flocculus

Based on embryologic comparative studies the flocculus was separated from the paraflocculus by Stroud (1895) and Smith (1903). This nomenclature was taken over by Ingvar (1918), Larsell (1937), and Dow (1935). The flocculus of the monkey contains 10 folia (Madigan and Carpenter 1971).

6.2.1 Connections

The intrinsic connections of the floccular cortex seem to be the same as for other parts of the cerebellum (Eccles et al. 1967). Information reaches the Purkinje cells (PC) via mossy fibers (MF) from various parts of the brainstem and via climbing fibers (CF) from the inferior olive. The only output is provided by axons of the Purkinje cells. Their activity therefore defines the output of the cortex. In the cat A. Brodal and Drahløs (1963) have described two different types of mossy fiber endings (rosettes) whose possibly different physiological functions have not yet been found. As in other parts of the cerebellum there is an overwhelming divergence and convergence of the MF input within the cortex. It has been estimated that the information of one mossy fiber (MF) distributes to about 20 000 PCs, and that one PC receives information from about 320 000 MFs (Ito 1984).

The densest MF projection to the flocculus arises from the vestibular nuclei and the perihypoglossal nuclei on both sides of the brainstem (rat, Blanks et al. 1983; cat, Sato et al. 1983a; Gerrits et al. 1984; monkey, Brodal and Brodal 1985; Rubertone and Haines 1981; Langer et al. 1985b). A. Brodal (1954) reported that in higher species the projections from the vestibular nuclei are far greater in number than those from the labyrinth. In the monkey the main origin of vestibulocerebellar fibers is from the central portion of the superior nucleus (SVN), the middle and caudal part of the medial nucleus (MVN), the descending nucleus (DVN), group y, and the interstitial nucleus of the vestibular nerve. No projection originates from the lateral vestibular nucleus (LVN).

Anatomical studies suggest a prominent projection of primary vestibular fibers to the flocculus, but physiological evidence is small (Ingvar 1918; Dow 1936; Brodal and Høivik 1964). In the monkey, after lesions of distinctive parts of the vestibular ganglia, degenerated fibers were found ipsilaterally in folia 5–10, mainly from ganglion cells innervating the macula (Carpenter et al. 1972). With neurochemical tracer techniques this projection of the eighth nerve has been confirmed by some authors (Carleton and Carpenter 1984; Rubertone and Haines 1981) and questioned by others (monkey, Langer et al. 1985b; cat, Korte and Mugnaini 1979). The existence of a large and functionally important projection of primary vestibular fibers has also been questioned on physiological ground because mossy fiber activity similar to that found in the vestibular nerve has not or rarely been recorded (monkey, Waespe et al. 1981; Lisberger and Fuchs 1978b; rat, Blanks and Precht 1983).

Other projections to the flocculus arise from the abducens nucleus (rat, Blanks et al. 1983; monkey, Brodal and Brodal 1985; Langer et al. 1985b), parts of the raphe reticular formation, the NRTP (cat, Gerrits et al. 1984), and from the contralateral rostral pontine nuclei (near the midline, from a region ventral to the NRTP, and from a dorsolateral region close to the inner aspect of the middle cerebellar peduncle; P. Brodal 1982; Langer et al. 1985b). In the monkey corticopontine fibers (P. Brodal 1978) and fibers from the pretectum (Benevento et al. 1977) end close to the dorsolateral parts of the pontine nuclei, which give rise to projections to the flocculus (P. Brodal 1978). The origin of climbing fibers to the flocculus has been described in the monkey by Brodal and Brodal (1982) and in the cat by Sato et al. (1983b). The main input is from the dorsal cap of Koy of the inferior olive and a minor input is from the medial accessory olive.

Efferent projections from the flocculus to the vestibular nuclei have been traced by Dow (1938), Angaut and Brodal (1967), and Haines (1977). Other projections are to the periphypoglossal nuclei (cat, Sato et al. 1982; Yingcharoen and Rinvik 1983), but the existence of this projection has explicitely been denied in the monkey (Balaban et al. 1981; Langer et al. 1985a). In the monkey the flocculus projects mainly to the vestibular nuclei, to a cell group termed the basal interstitial nucleus of the cerebellum (Langer 1985), and to cells near and in the y-group (Langer et al. 1985a). The projection is strictly ipsilateral

(Haines 1977). The projection to the vestibular nuclei, traced by autoradiography and HRP (horseradish peroxidase), is mainly to the central core of the SVN, to the rostral part of the MVN, and to the medial part of the DTN. There are no projections to the LVN. The first of the two most important aspects are that the distribution of labelled cells in the MVN is nonhomogenous, suggesting that only certain cell populations within the MVN receive an input from the flocculus. The second aspect is that the floccular output to the vestibular nuclei seems to be organized in small longitudinal zones (cat, Sato et al. 1982; monkey, Balaban et al. 1981). The MVN receives its main input from a middle small zone of 500–750 μm in width which extends over all folia of the flocculus (see also Carleton and Carpenter 1983). This zone seems to be functionally involved mainly with horizontal eye movements (Balaban and Watanabe 1984). Electrical stimulation in a central region of the flocculus elicited horizontal nystagmus with slow phases to the ipsilateral side (Ron and Robinson 1973). The SVN receives projections from two zones just lateral and medial to this middle zone, having a width of 300–500 μm.

6.2.2 Flocculectomy

The effects of uni- and bilateral surgical ablation of the flocculus have been described in the rabbit (Koehn et al. 1981; Honrubia et al. 1982; Ito et al. 1982; Barmack and Petterossi 1985), cat (Precht et al. 1985), and monkey (Takemori and Cohen 1974; Zee et al. 1981; Waespe et al. 1983). Usually large parts of the paraflocculus were also removed together with the flocculus.

Effects on Eye Movements. After flocculectomy, a prominent finding in all monkeys was a gaze-paretic nystagmus in the horizontal plane. Monkeys were unable to hold eccentric eye positions in darkness. The eyes slowly drifted back from a lateral eye position to the midposition with a time constant between 1–3 s (Zee et al. 1981; Waespe et al. 1983). Saccades were normal in all their parameters.

Vestibular nystagmus was not much affected (Fig. 21, dashed lines). Gain could increase or decrease and changes were always less than 30% relative to the preoperative values. In seven monkeys (three from the study of Waespe et al. 1983 and four from the study of Zee et al. 1981) the preoperative gain was 0.88 (range 0.82–0.97) and the postoperative gain was 0.86 on average (range 0.63–1.15). The time constant of vestibular nystagmus decay could be slightly less postoperatively, but still remained in the range found in normal, habituated monkeys.

The strongest effect on *optokinetic nystagmus* was seen at high stimulus velocities. Nystagmus gain was then always reduced, whereas it was not much affected for stimulus velocities up to about $60^{\circ} \cdot s^{-1}$ (Fig. 20A, B; Table 1). Postoperatively the highest OKN velocities corresponded to the preoperative OKAN-saturation velocities (Table 1). The transition of OKN to OKAN was smooth and did not show the fast decay in eye velocity so characteristic in the normal monkey. Time constants of OKAN were in the range seen in normal, habituated monkeys (Table 1). The initial fast increase in eye velocity at the onset of the OKN was strongly reduced (Fig. 20C; Table 1). Consistent with the much reduced rapid initial response to steps in surround velocity, monkeys were unable to follow accelerations of the visual surround with eye accelerations faster than $3-5^{\circ} \cdot s^{-2}$. They could, however, decelerate their eyes considerably faster (Waespe et al. 1983).

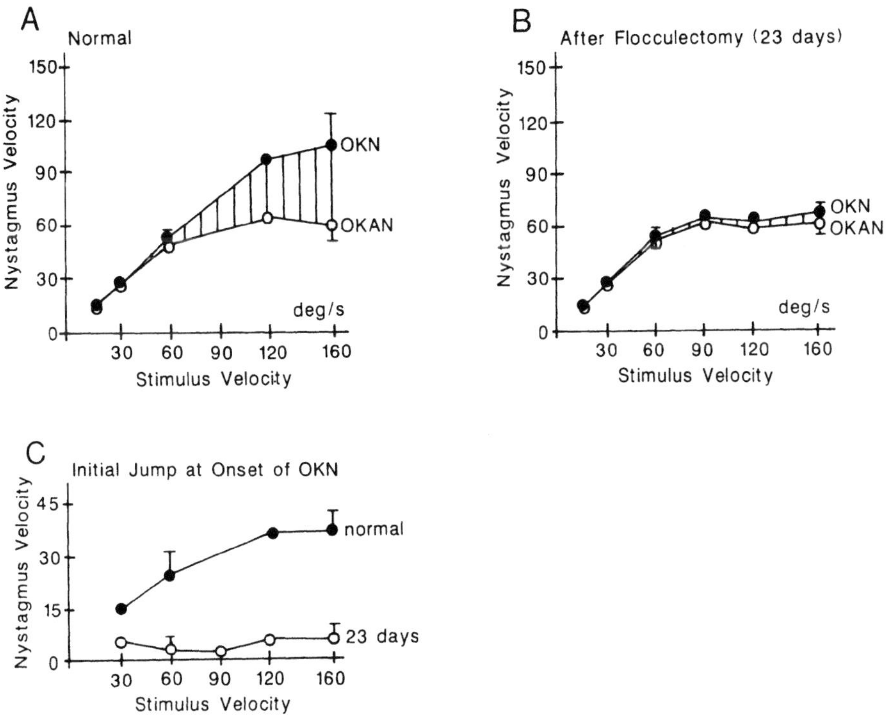

Fig. 20A, B. Steady state velocity of OKN and initial velocity of OKAN in a monkey before **(A)** and 23 days after **(B)** bilateral complete flocculectomy (and partial paraflocculectomy). **C** Initial fast increase of OKN. All values are averages of at least six trials, SDs are given only for stimulus velocities of 60 and $160^{\circ} \cdot s^{-1}$. The *vertically hatched area* in **A** and **B** indicates the amount of the fast decay in eye velocity during the transition of OKN to OKAN

Table 1. Values of OKN and OKAN parameters before and after bilateral flocculectomy. All values in deg/s, except time constant of OKAN [s]

	OKN				OKAN			
	Initial jump		Steady state		Initial velocity		Time constant	
	pre	post	pre	post	pre	post	pre	post
$60^{o} \cdot s^{-1}$	33 (22–42) n = 7	15 (3–21) n = 7	58 (54–60) n = 7	52 (42–58) n = 7	49 (48–50) n = 3	50 (47–53) n = 3	36 s (14–59) n = 7	31 s (13–52) n = 7
$120^{o} \cdot s^{-1}$	53 (37–82) n = 3	11 (6–15) n = 3	101 (97–109) n = 3	55 (51–63) n = 3	61 (59–66) n = 3	47 (48–59) n = 3	–	–

n = 7 (4 monkeys from Zee et al. 1981; 3 monkeys from Waespe et al. 1983)
n = 3 (monkeys from Waespe et al. 1983)

In other species effects of bilateral flocculectomy have been quite similar except that the gain of vestibular nystagmus has been reported always to be lower postoperatively.

In the rabbit lesions of the NRTP, believed to send visual information to the flocculus via MFs (Maekawa et al. 1981a, b), strongly reduced the gain of OKN induced by stimulation of the contralateral eye (Miyashita et al. 1980). Lesions of fibers from the pretectum to the dorsal cap of Koy of the inferior olive did not change the gain of OKN.

Unilateral flocculectomy had similar effects on OKN, OKAN, and VN as a bilateral flocculectomy. Effects were stronger on OKN with slow phases to the ipsilateral than contralateral side (Waespe et al. 1983). In the first week after the operation, there was a strong spontaneous nystagmus in darkness to the ipsilateral side which was compensated within 7–10 days (cat, Flandrin et al. 1983). Gain of VN, after compensation of the spontaneous nystagmus, had not changed significantly.

Smooth pursuit (SP) was also affected by bilateral flocculectomy. For target velocities between 25–40$^{\circ} \cdot s^{-1}$, gain dropped from 0.95 to 0.64. SP was similarly impaired for all velocities between 10 and 60$^{\circ} \cdot s^{-1}$. It is important to note that SP was not only affected in the high velocity range like OKN. With time SP eye movements improved and the gain of SP could reach a value of 0.78 several months postoperatively (Zee et al. 1981). There is some discrepancy between the gain reduction of SP of about 30% and the stronger gain reduction of the initial fast increase in OKN of about 50% (at 60$^{\circ} \cdot s^{-1}$ stimulus velocity, Zee et al. 1981). This discrepancy questions the theory that the initial fast increase of OKN is generated by the SP system (see Sect. 2.7).

Visual suppression of VN was also affected by flocculectomy. When presented with a monkey-stationary visual surround during animal rotation (conflict-stimulation), monkeys had difficulties to suppress or attenuate peak VN velocities postoperatively. The ability of the monkey to attenuate VN was a function of the amount of the acceleration applied. This is demonstrated in Fig. 21. In response to a velocity step the monkey was unable to attenuate peak nystagmus velocity (Fig. 21A, B). In two monkeys peak nystagmus velocity was attenuated during conflict-stimulation in response to a step in head velocity preoperatively by 64% (SD 8.5%) and postoperatively by only 20% (SD 11.5%); however the time constant of nystagmus decay remained similar. Whereas the decay time constant was 27.5 s in darkness, it was reduced to 4.0 s (SD 0.8) preoperatively and to 5.7 s (SD 0.6) post-

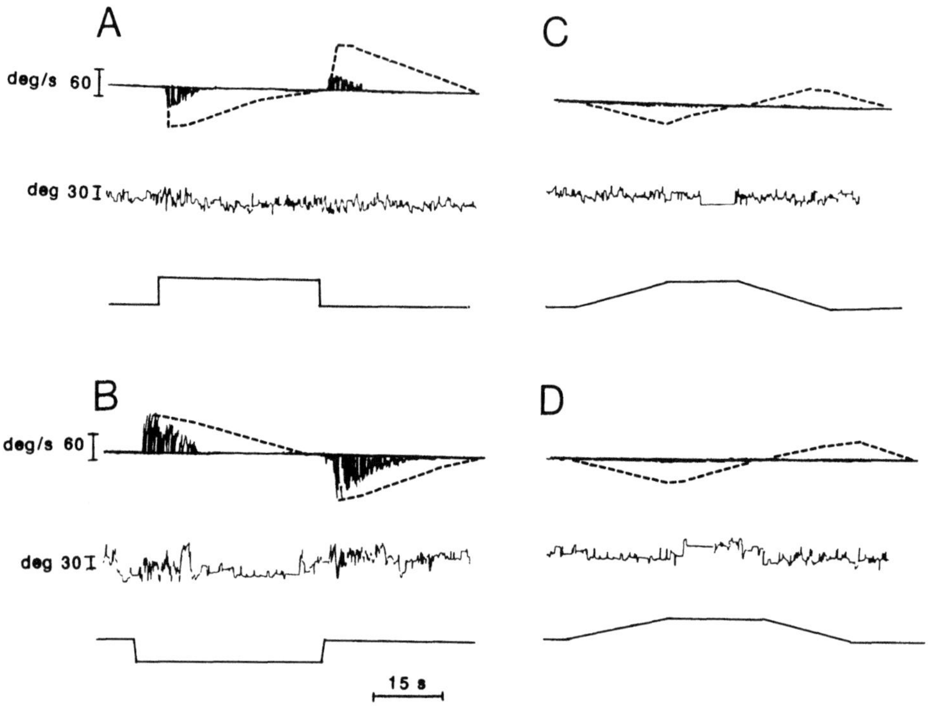

Fig. 21 A–D. Velocity of slow phases of vestibular nystagmus during conflict stimulation before (**A, C**) and after (**B, D**) bilateral flocculectomy. *First trace,* eye velocity; *second trace,* horizontal eye position; t*Third trace,* velocity profile of stimulus. In each subfigure the *dashed line* represents the VN velocity during control trials in darkness. In **A, B** a velocity step and in **C, D** a constant acceleration of $5^{\circ}\ s^{-2}$ (constant velocity $120^{\circ}{\cdot}s^{-1}$) is applied

operatively during conflict-stimulation. In contrast, with a lower accleration of $5^{\circ}{\cdot}s^{-2}$ monkeys had no difficulties in suppressing vestibular nystagmus even postoperatively (Fig. 21C, D). During sinusoidal conflict-stimulation (at 0.25 Hz) the percentage suppression of VN velocity dropped from a preoperative average of 51% to a postoperative value of 20% (Zee et al. 1981). If the monkey was rotated with a single head-fixed target light instead of the full field, he also had difficulties in suppressing the VOR postoperatively (Zee et al. 1981).

Presentation of a monkey-stationary visual surround for short periods during postrotatory nystagmus or OKAN strongly reduces nystagmus velocity during the period of fixation and also reduces the occurrence of nystagmus after the light is turned off again (Sect. 2.6.2). This has been called "dumping". It was little affected by flocculectomy. The "dumping" time constant was 3.5 s for VN and 2.8 s for OKAN after flocculectomy and thus similar to that in the normal monkey (Waespe et al. 1983).

Activity of Vestibular Neurons in the Vestibular Nuclei. After cerebellectomy, Keller and Precht (1979) found in the paralyzed, alert cat qualitatively unchanged visual responses of vestibular nuclei neurons. Neurons were, however, saturated at low stimulus velocities of about $10^{\circ} \cdot s^{-1}$ and the strength of modulation was reduced compared to the normal cat. In agreement with these single cell recordings, OKN in cerebellectomized cats was normal up to $10^{\circ} \cdot s^{-1}$ stimulus velocity, but OKN was extremely velocity limited at higher stimulus speeds. Postrotatory nystagmus could still be suppressed in the light. Godaux and Vanderkelen (1984) reported similar findings. During sinusoidal stimulation OKN could not be elicited at frequencies above 0.15 Hz, whereas OKN was only moderately affected during constant stimulus velocity of $60^{\circ} \cdot s^{-1}$. This suggests that after cerebellectomy OKN is not only velocity limited, but even more so acceleration limited. The VOR could not be suppressed visually for frequencies of rotation above 0.25 Hz. The effects of hemicerebellectomy on eye movements in the monkey have been reported by Westheimer and Blair (1974) and in man by Estanol et al. (1979).

After bilateral flocculectomy in the monkey, neurons in the vestibular nuclei receiving an input from the horizontal canals were still modulated during OKN and OKAN, qualitatively similarly to those in normal monkeys (Waespe and Cohen 1983). On average the strength of modula-

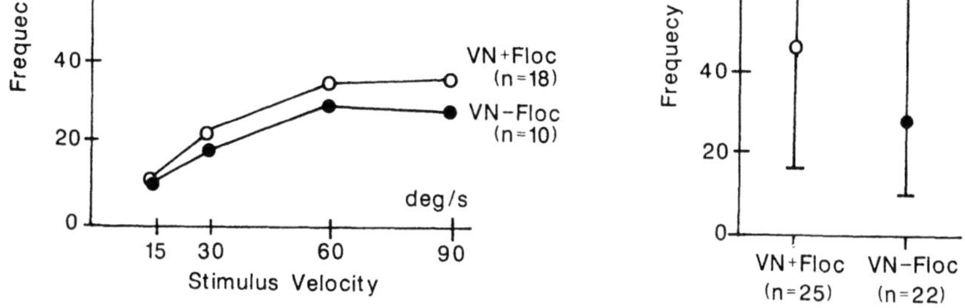

Fig. 22A. Comparison of firing rate changes of type I and type II neurons in the vestibular nuclei in normal (*VN + Floc*) and flocculectomized monkeys (*VN – Floc*) during different constant velocities of optokinetic stimulation. **B** Comparison of firing rate changes induced in type I neurons during optokinetic stimulation at $60^{\circ} \cdot s^{-1}$ in normal and flocculectomized monkeys. *Open circles* are mean values and *vertical bars* indicate the range of maximal and minimal frequency changes (Waespe and Cohen 1983)

tion was sligthly less than normal (Fig. 22). The time course of OKAN and VN was also reflected in the modulation of neuronal activity similar to that in the normal monkey. Furthermore, even after flocculectomy, during conflict-stimulation neuronal activity decayed with a time constant much shorter than after vestibular stimulation in darkness. This demonstrates that full-field stimulation can still influence activity of vestibular nuclei neurons in the absence of pathways through the flocculus.

6.2.3 Neuronal Activity in Normal and Vestibular Neurectomized Monkeys

Lesions of the flocculus (usually combined with partial lesions of the paraflocculus) suggest that these structures are involved in the "direct" visual-oculomotor pathways (Sect. 2.7). After flocculectomy, fast changes in eye velocity during OKN are severely reduced, whereas slow changes and OKAN are preserved. Smooth pursuit is also reduced. Single cell recordings in normal and vestibular neurectomized monkeys indeed demonstrate the functional importance of the flocculus in mediating eye velocity and acceleration command signals for smooth pursuit eye movements and for those components of OKN which are attributable to activity in "direct" visual-oculomotor pathways.

Purkinje Cell Activity. The activity of Purkinje cells (PC) is characterized by simple spikes (SS) and complex spikes (CS). The SS activity, occuring spontaneously at a rate of about 80 imp$\cdot$s^{-1} (range 20–140) is assumed to be modulated by the MF input, and the CS activity, occurring at a rate of 1–2 imp$\cdot$s^{-1}, is assumed to be evoked and modulated by the CF input.

OKN/OKAN. In normal monkeys (Waespe and Henn 1981; Büttner and Waespe 1984; Waespe et al. 1985b) no PC was found to be consistently modulated during steady state OKN if the stimulus velocity was below 30–60$^{\circ}\cdot$s^{-1}. Only with eye velocity above 20–50$^{\circ}\cdot$s^{-1} were some PCs activated (Figs. 23B, 25A). The threshold eye velocity at which PCs started to be modulated ranged between 17–52$^{\circ}\cdot$s^{-1} with an average of 33$^{\circ}\cdot$s^{-1}. They were also not found to be modulated during OKAN. The threshold corresponded approximately to the OKAN saturation velocity (Fig. 25A). For eye velocities above 20–50$^{\circ}\cdot$s^{-1} firing rate changes and eye velocities were linearly related and eye velocity sensitivity ranged between 0.52 and 2.95 imp$\cdot$s^{-1} per $^{\circ}\cdot$s^{-1} (average 1.45). When the drum was slowly decelerated with 5$^{\circ}\cdot$s^{-2} from a velocity of 120$^{\circ}\cdot$s^{-1} resting discharge activity was reached at an instantaneous eye velocity of OKN between 30 and 50$^{\circ}\cdot$s^{-1}. When eye velocity slowed down during constant velocity stimulation, neuronal modulation also decayed.

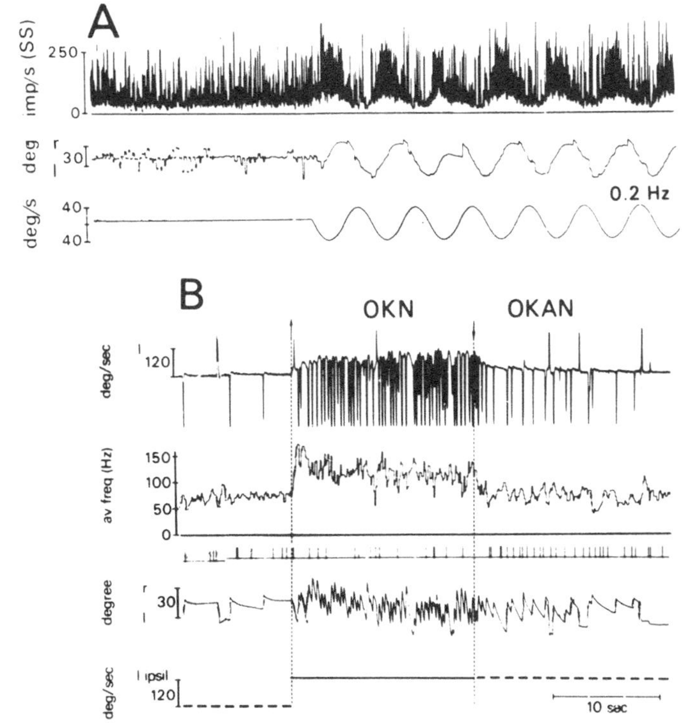

Fig. 23A. Instantaneous SS activity (*first trace*) of a PC in a normal monkey during sinusoidal SP. *Second trace,* horizontal eye position; *third trace,* velocity of target light. **B** SS activity of another PC during OKN at a stimulus velocity of $120^{\circ}\cdot s^{-1}$ and during OKAN. *First trace,* eye velocity; *second trace,* average SS activity; *third trace,* CS activity; *fourth trace,* eye position; *fifth trace,* stimulus velocity

Over 90% of the PCs increased firing rate for eye velocities to the ipsilateral side. All PCs modulated during OKN were also modulated during conflict-stimulation, i.e., attenuation of VN in the presence of a monkey-stationary visual surround. Transient changes in SS activity were observed initially at the onset of OKN during the fast increase in eye velocity and when the drum was stopped in the light (Fig. 24A). In response to sudden illumination of the moving drum, the frequency changes within the first 100 ms contained a component related to eye acceleration and another component related to eye velocity. Average eye acceleration sensitivity was $0.25\ \text{imp}\cdot s^{-1}$ per $^{\circ}\cdot s^{-2}$.

After bilateral vestibular neurectomy OKAN is permanently abolished (Sect. 3.5). One can interpret this as an OKAN saturation velocity of zero. The eye velocity threshold at which PCs started to respond was then on average $0.5^{\circ}\cdot s^{-1}$ (range $0-10^{\circ}\cdot s^{-1}$ in the ipsilateral direction and $0-12^{\circ}\cdot s^{-1}$ in the contralateral direction; Fig. 25B). After neurectomy PCs

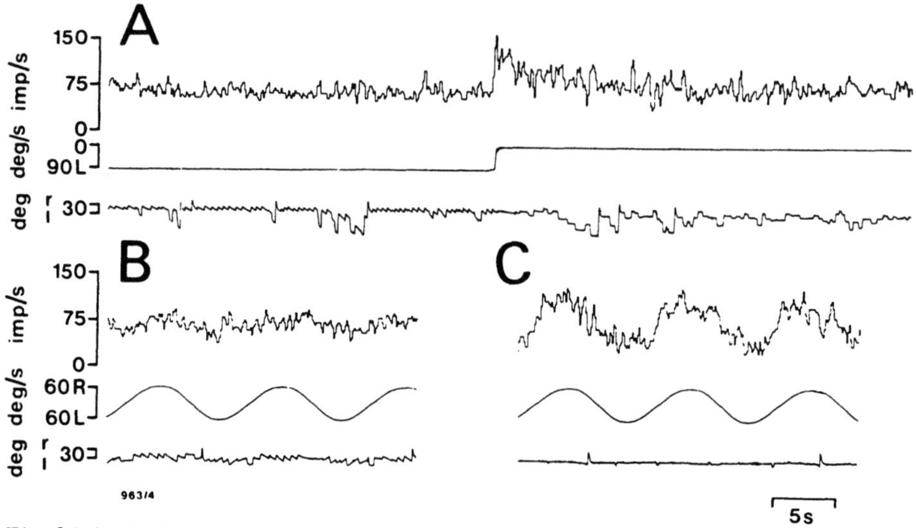

Fig. 24 A–C. Average SS activity of a PC during optokinetic stimulation at $90^{\circ} \cdot s^{-1}$, and in response to sudden deceleration of the full field in the light (**A**), during sinusoidal rotation of the monkey (VOR, **B**), and VOR-suppression (**C**). *First trace,* SS activity; *second trace,* stimulus velocity; *third trace,* horizontal eye position (Büttner and Waespe 1984)

were thus modulated during steady-state OKN over ranges of eye velocities that do not cause a response in a normal monkey (Fig. 25B; Waespe et al. 1985b). Sensitivity to eye velocity ranged between 0.3 and 3.1 $imp \cdot s^{-1}$ per $^{\circ} \cdot s^{-1}$ (average 1.0). Again over 90% of the PCs increased firing rate of SS activity with eye velocity to the ipsilateral side and decreased it with eye velocity to the contralateral side.

Smooth Pursuit. During SP a group of PCs is modulated in relation to eye velocity with an average sensitivity between 0.8 and 1.14 $imp \cdot s^{-1}$ per $^{\circ} \cdot s^{-1}$ (Fig. 23A; Lisberger and Fuchs 1978a; Miles et al. 1980a, b; Büttner and Waespe 1984). The same PCs are also modulated during VOR-suppression in relation to head velocity (Fig. 24C). In over 90% of the PCs, SS activity increased with eye velocity during SP or with head velocity during VOR-suppression to the ipsilateral side and decreased with eye or head velocity to the contralateral side. In the study of Miles and coworkers (1980b) there was an average phase lead at 0.2 Hz ($\pm 20^{\circ}$) between peak firing rate and peak eye velocity of 7.6° (range 59° lead to 53° phase lag). About 70% of these PCs showed some eye position component during maintained eye fixation with an average sensitivity of 0.26 $imp \cdot s^{-1}$ per $^{\circ}$ (see below). The threshold eye velocity is probably $0^{\circ} \cdot s^{-1}$ as PCs are already modulated at SP velocities of $8^{\circ} \cdot s^{-1}$ (Fig. 25A, right side; Miles et al. 1980). We found no indication that

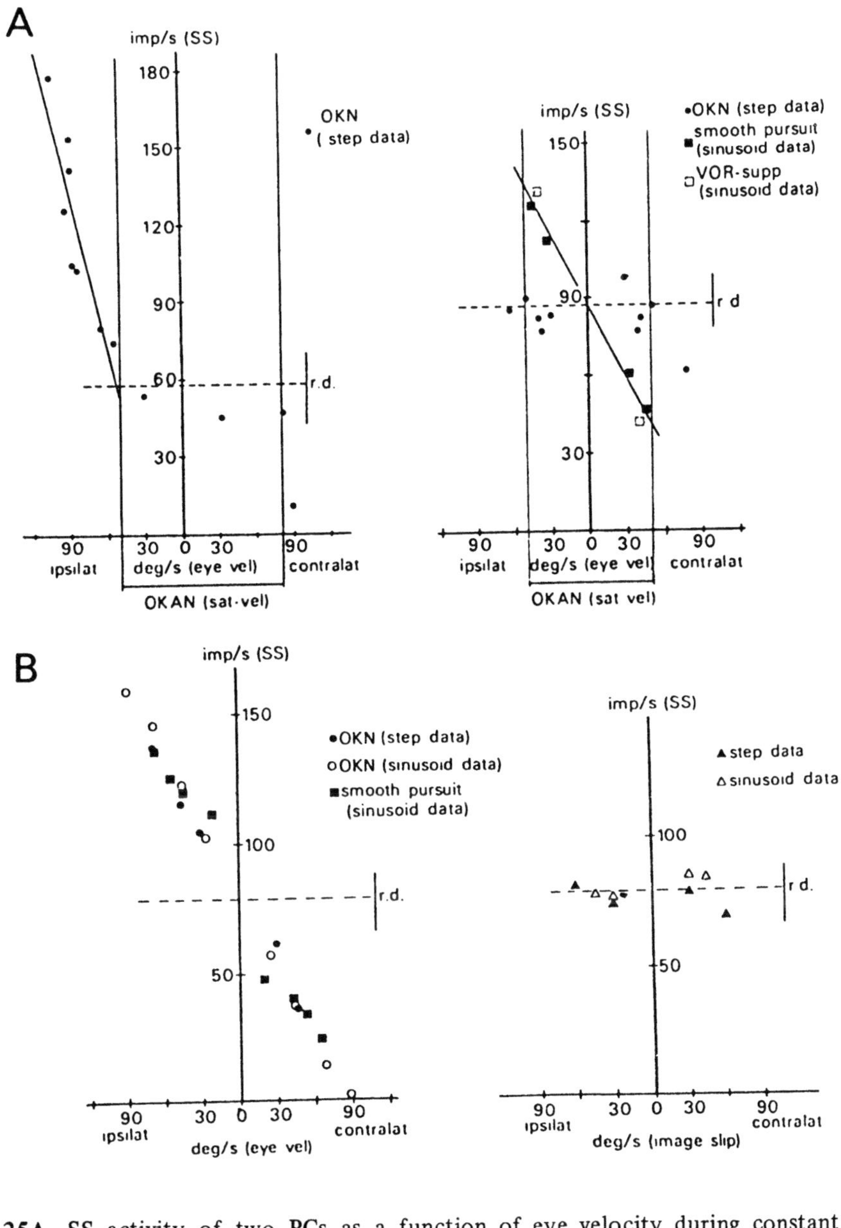

Fig. 25A. SS activity of two PCs as a function of eye velocity during constant velocity OKN (*left side*) and during constant velocity OKN, sinusoidal SP, and as a function of head velocity during VOR-suppression(*right side*) in two normal monkeys. *Symbols* are averaged data, the saturation velocities of OKAN to the ipsilateral and contralateral side are indicated by vertical lines and at the bottom below the abscissa. Regression lines are drawn on the left side for values above the saturation velocity and on the right side for values during SP assuming a threshold velocity of $0^{o}\cdot s^{-1}$. **B** SS activity of a PC as a function of eye velocity (*left side*) or image slip velocity (*right side*) in a monkey after bilateral vestibular neurectomy. SS modulation is dependent only on eye velocity regardless of how it is induced, either by a full field as during OKN or by a single target light as during SP. In **A** and **B** the *horizontal dashed line* indicates the average resting discharge activity and the *vertical bar* at the right side its range (adapted from Waespe et al. 1985b)

PCs modulated during SP and OKN form different subgroups (Büttner and Waespe 1984). Thus, SS activity of PCs is modulated during steady state OKN only at eye velocities above 20–50$^{o}\cdot s^{-1}$ but during SP at any eye velocity. During transient changes of triangle-wave tracking, over 50% of PCs had also a transient over- or undershoot in SS activity suggesting that the firing rate was also related to eye acceleration (Miles et al. 1980; Lisberger and Fuchs 1978a).

As after bilateral vestibular neurectomy SP eye movements are intact, it is not expected that PC modulation is affected in any way by the operation. In bilaterally neurectomized monkeys the modulation of PCs during SP is indeed unchanged compared to normal monkeys (Waespe et al. 1985b). Sensitivity to eye velocity is on average 1.28 imp$\cdot s^{-1}$ per $^{o}\cdot s^{-1}$. For P-cells tested during SP and OKN eye velocity sensitivity is comparable regardless of how eye velocity is induced (Fig. 25B). Eye velocity sensitivity was 1.23 imp$\cdot s^{-1}$ per $^{o}\cdot s^{-1}$ during sinusoidal SP compared to 1.11 imp$\cdot s^{-1}$ per $^{o}\cdot s^{-1}$ during sinusoidal OKN.

These results strongly support the notion that eye velocity during SP and OKN is not coded by different subgroups of PCs. However, they suggest that eye velocity is coded during steady-state OKN only for values higher than 20–50$^{o}\cdot s^{-1}$ in a normal monkey.

Image Slip Velocity. With few exceptions PCs responsive to OKN and SP eye velocity were not modulated by full-field image slip (Fig. 25B, right side). This is apparent during OKN when eye velocity considerably lags behind stimulus velocity, and during OKN-suppression. Also during SP, firing rate changes are related to eye velocity and not to target velocity (Lisberger and Fuchs 1978a).

VOR and VOR-Suppression. About 50% of the PCs modulated during SP and/or OKN were not modulated during VOR in darkness. If there was some modulation, it was weak and much less than during SP or during VOR-suppression (Fig. 24B, C). In the study of Miles and coworkers (1980) head velocity sensitivity of P-cells was 0.10 imp$\cdot s^{-1}$ per $^{o}\cdot s^{-1}$ during VOR (0.2 Hz, ± 20 deg) compared to 1.19 during VOR-suppression. During VOR-suppression – as during SP – over 90% increased firing rate with head velocity to the ipsilateral side. On average the strength of modulation of P-cells during VOR-suppression and SP is comparable (Lisberger and Fuchs 1978a; Miles et al. 1980b; Büttner and Waespe 1984). Average sensitivity to "head velocity" during VOR suppression ranged between 0.88 and 1.12 imp$\cdot s^{-1}$ per $^{o}\cdot s^{-1}$. At 0.2 Hz (± 20°) peak firing rate led peak "head velocity" by an average of 11.6^{o} (range 55° lead to 57° lag; Miles et al. 1980b).

PC activity has not been recorded in the cat. In normal rabbits and after brainstem lesions PCs were found to be mainly modulated by image slip velocity. The relative contribution of visual signals to PC modulation dominated over that of eye velocity and proprioceptive signals during OKN (Miyashita 1984; Miyashita and Nagao 1984).

Eye Position Sensitivity. Dependent on the study cited, about 20%–70% of the PCs modulated by eye velocity during SP showed also some sensitivity to different eye positions during maintained fixation. In addition, SS activity of most PCs was also loosely modulated by fast eye movements (Lisberger and Fuchs 1978a; Miles et al. 1980b; Waespe et al. 1985b). Other PCs showed a much stronger and reproducible relation to eye positions and fast eye movements (Noda and Suzuki 1979a, b). Discharges of 13% of all floccular PCs changed tonically with shifts of gaze. This position component was also observable during SP, but only at low velocities (Noda and Warabi 1982).

Complex Spike Activity. Usually, CS activity was found not to be modulated during VOR, VOR-suppression, and SP. During conflict stimulation, when nystagmus was not completely suppressed, and during OKN, CS activity was modulated. This modulation was always reciprocal to the modulation of SS activity. We had the impression that the CS activity was modulated whenever image slip was present. A clear dependence on its velocity could not be established. During steady-state OKN at $30^{\circ}\cdot s^{-1}$ with only a small image slip and at $160^{\circ}\cdot s^{-1}$ stimulus velocity with an average image slip of $70^{\circ}\cdot s^{-1}$, CS activity was similarly modulated by about 30% of its rest levels (Waespe et al. 1981).

Mossy Fiber Input Activity. Input activity can be recorded from MFs and most probably also from glomeruli. MF input activity to the flocculus is diverse, transmitting oculomotor, vestibular, and visual signals. Some MF units show saccade-related burst and position-related intersaccadic tonic activity. They exhibit regular firing rates that increase for eye positions in a preferred direction and exhibit a burst of spikes in the preferred direction and cease firing for saccades in the opposite direction (Fig. 26A). During smooth pursuit many of these MF units also code eye velocity, and the peak firing rate shows a phase lead in relation to eye position (Fig. 26B; Lisberger and Fuchs 1978b; Miles et al. 1980b; Noda and Warabi 1982). Some units predominantly reflect eye velocity and other units eye position. In the study by Miles and coworkers (1980b) the mean eye position sensitivity of burst-tonic units was 2.51 $imp\cdot s^{-1}$ per $^{\circ}$. During SP at 0.2 Hz ($\pm 20^{\circ}$), peak firing rate lagged peak eye velocity by 61.1° (range 2°–91°). Other

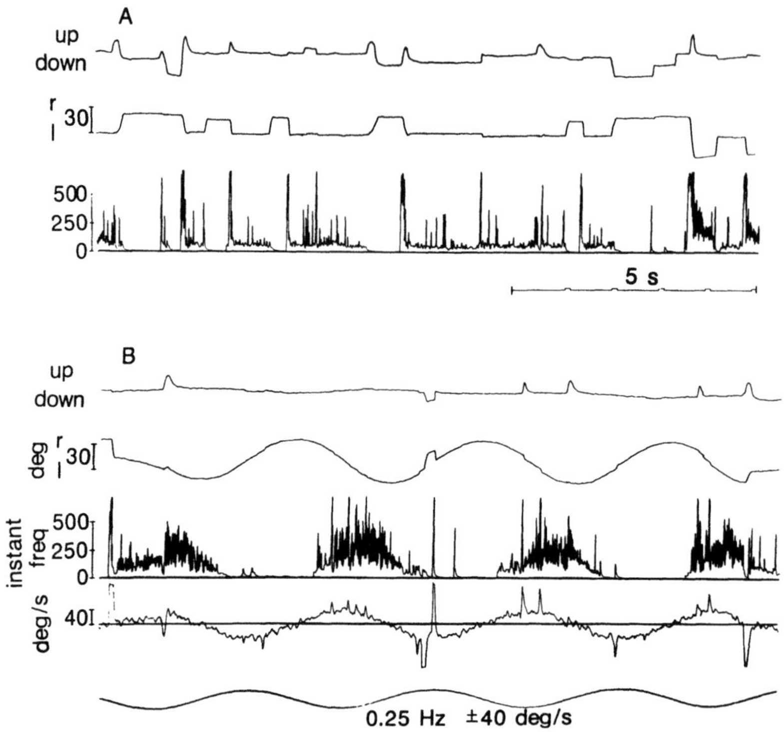

Fig. 26A, B. Firing rate of a burst-tonic MF input neuron during spontaneous eye movements (**A**) and during SP (**B**). *First trace,* vertical eye position; *second trace;* horizontal eye position; *third trace,* instantaneous firing rate, and in **B** in addition horizontal eye velocity and target velocity. For saccades to the left, there is a burst in activity. Tonic activity increases for eye positions to the left. In **B** during SP, peak firing rate occurs approximately between peak eye velocity and eye position to the left.

MF units exhibited bursts of spikes for saccades only (saccade only units); other units were sensitive to different steady-state eye positions without exhibiting saccade-related activity changes (tonic units).

MF units modulated by vestibular stimulation usually also showed activity changes related to fast eye movements and different eye positions (Lisberger and Fuchs 1978; Miles et al. 1980; Waespe et al. 1981). These units are assumed to be vestibular units as their activity is still modulated during conflict stimulation when VN is suppressed (Fig. 27A, B, see below). These vestibular MF units were either vestibular-only units, vestibular-plus-saccade units, and vestibular-plus-position units. In the vestibular-plus-position units, the eye position sensitivity was 1.03 imp$\cdot$s^{-1} per ^{o} (range 0.15 to 3.5, Miles et al. 1980), and "head" velocity sensitivity at 0.2 Hz (± 20°) was 1.30 imp$\cdot$s^{-1} per $^{o}\cdot$s^{-1} (Fig. 27). During VOR suppression the phase lead was on average 14.1^{o} relative to "head" velocity. During SP at 0.2 Hz (± 20°) peak firing rate led peak eye

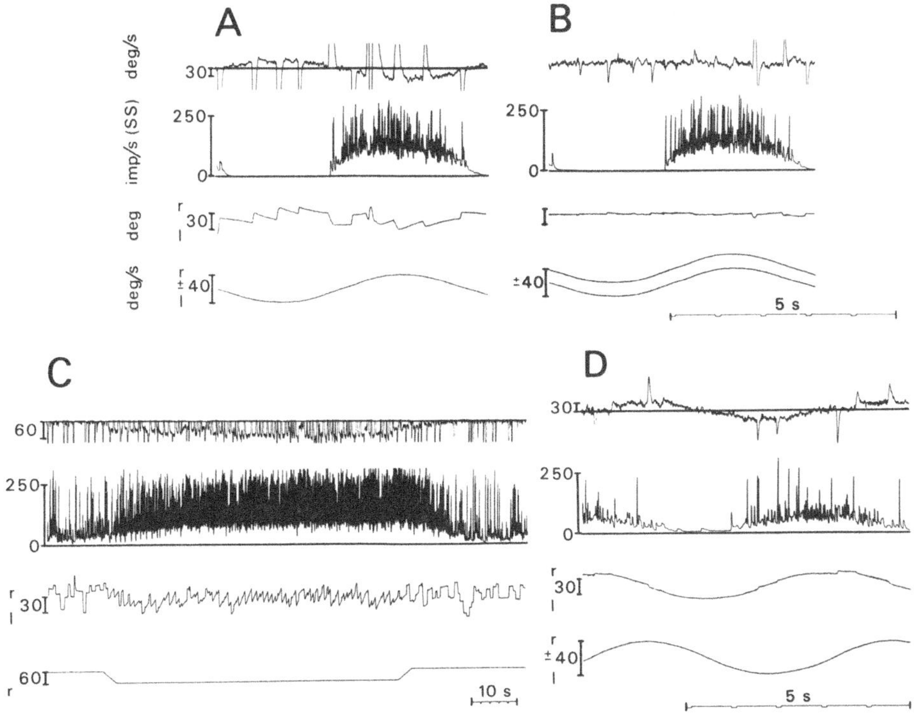

Fig. 27 A–D. Firing rate of a vestibular MF input neuron during VOR in the dark (**A**), VOR-suppression (**B**), optokinetic nystagmus (**C**), and during SP (**D**). ***First trace,*** eye velocity; ***second trace,*** instantaneous firing rate; ***third trace,*** horizontal eye position; ***fourth trace,*** turnable velocity (**A**) and head and target velocity (**B**, 0.2 Hz, ± 40 $^{\circ}\cdot s^{-1}$ peak velocity), or optokinetic stimulus velocity (**C**, 60$^{\circ}\cdot s^{-1}$), and target velocity (**D**, 0.2 Hz, ± 40$^{\circ}\cdot s^{-1}$). Note different time base in **C**. In **A** and **C** the modulation increases for slow eye movements to the same (right) side. During SP modulation is mainly in phase with eye position (Waespe et al. 1985b)

position, suggesting the existence of an eye velocity sensitivity component. Phase lag relative to eye velocity was 60° and eye velocity sensitivity was 0.46 imp$\cdot s^{-1}$ per $^{\circ}\cdot s^{-1}$ (range 0.07 to 1.83). In our study (Waespe et al. 1981) 91% of MF units responsive to vestibular stimulation showed also a modulation during OKN, OKAN, and conflict-stimulation (Fig. 27). During OKN, firing rate changes increased monotonically up to stimulus velocities of 40–60$^{\circ}\cdot s^{-1}$. During OKN-suppression, units were also modulated in relation to image slip velocity, but the strength of modulation was greater when OKN was allowed to occur (Fig. 28B). Modulation was bidirectional, i.e., increased for optokinetic stimulation in one direction and decreased for stimulation to the opposite side. Most importantly, the firing rate increased for vestibular and optokinetic stimulation for slow phase velocities of nystagmus in the same direction (Fig. 27A, C). During OKAN firing rate was related to eye velocity.

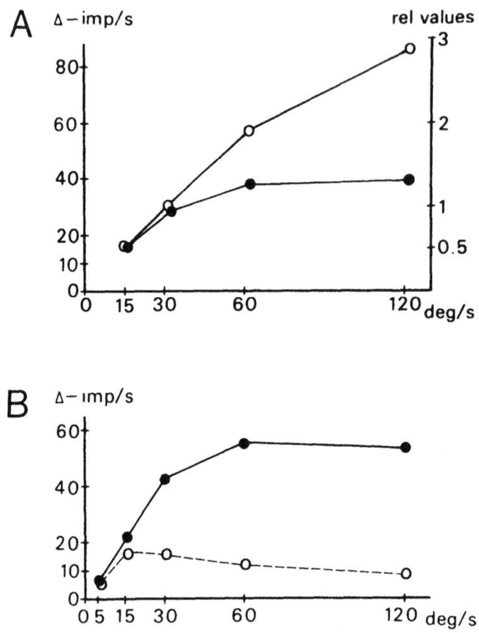

Fig. 28A. Changes in firing rate (*filled circles, left ordinate*) of vestibular MF input units and steady-state velocities of OKN (*open circles, right ordinate,* relative values) as a function of different optokinetic stimulus velocities. Averages for 18 units. **B** Changes of firing rates of 7 units during OKN (*solid line*) and during OKN suppression (*dashed line*). Units are also modulated in relation to image slip velocity but modulation is less when OKN is suppressed (Waespe et al. 1981)

Of 62 vestibular MF units, 23% exhibited some sensitivity to different eye positions. These units were also modulated during SP with an eye position and eye velocity component. All other units were not modulated during SP. The eye position sensitivity of the 14 vestibular-plus-eye- position units was 0.68 imp$\cdot s^{-1}$ per $^{\circ}$ and the sensitivity to eye velocity during SP was 0.23 imp$\cdot s^{-1}$ per $^{\circ}\cdot s^{-1}$. However, the same units showed an eye velocity sensitivity during OKN (at 30 and $60^{\circ}\cdot s^{-1}$ stimulus velocity) of 1.40 imp$\cdot s^{-1}$ per $^{\circ}\cdot s^{-1}$. Thus, the same units showed quite different eye velocity sensitivities depending on how eye velocity is induced, i.e., by a full field or by a single target light. During VOR the "head" velocity sensitivity was 1.44 imp$\cdot s^{-1}$ per $^{\circ}\cdot s^{-1}$ and was slightly less during VOR-suppression. The distribution of type I and type II vestibular MF units was about equal and the units showed identical features.

Thus, over 90% of units carrying a vestibular or "head" velocity signal to the flocculus showed features which were never observed in primary vestibular neurons. We have suggested that these MF units originate in neurons of the vestibular nuclei, which we named "central vestibular neurons".

Considerable controversy remains over the existence of a separate visual MF input to the flocculus. Activity which showed some relation to visual stimuli, sometimes coding image slip velocity of a full field, was recorded, albeit rarely (Noda 1981; Miles et al. 1980b; Waespe et al. 1981).

6.2.4 *Conclusions*

Much evidence has accumulated supporting the hypothesis that the two dynamically different components, the slow and the fast component, in the OKN slow phase velocity response (Sect. 2.7) depend on different neural populations. The time course of changes in firing rate of floccular PCs in normal and vestibular neurectomized monkeys closely follows the time course and velocity of the fast component. Firing rates of floccular PCs do not reflect sensory information about image slip but rather mediate an eye velocity and acceleration command signal. Where and how the sensory information is transformed into the oculomotor command signal is presently unknown. There is strong evidence that one of the main functions of the flocculus in primates may be the computation of a signal to induce or suppress following eye movements during visual-vestibular interaction. The "vestibular" signal transmitted to the flocculus via MFs is complex and reflects the amount to which a subgroup of neurons in the vestibular nuclei (called "central vestibular neurons"; Sect. 5.1) contribute to eye velocity during head movements, during suppression of VOR, and during OKN and OKAN. The modulation of floccular PCs can only be interpreted together with the modulation of "central vestibular neurons". Whenever the signal carried by these neurons is inappropriate to induce or suppress eye movements for stable vision, floccular PCs and "central vestibular neurons" behave complementarily to each other during visual-vestibular interaction. A major question concerns where in the brainstem the target cells of the floccular PCs are located. According to anatomical studies in the monkey, they can only be located within the vestibular nuclei complex. These target cells probably combine signals of the "central vestibular neurons" and of the floccular PCs, or the signals of the "indirect" and "direct" visual-oculomotor pathways.

6.3 Nodulus and Ventral Uvula

The nodulus and ventral uvula (folia X and IX) form the vermian part of the vestibulo-cerebellum. Primary and secondary vestibular neurons abundantly project to this part of the vestibulo-cerebellum (rabbit, Epema et al. 1985; cat, Korte and Mugnaini 1979; monkey, Carpenter et al. 1972; Rubertone and Haines 1981). There is general agreement that the fastigial nuclei do not get a direct input from primary vestibular neurons, although fastigial neurons are sensitive to rotational stimuli (monkey, Gardner and Fuchs 1975; Kimm et al. 1976). The nodulus and ventral uvula project to the fastigial nuclei and also directly to the vestibular nuclei (Dow 1938). Target cells in the vestibular nuclei are different from those of the flocculus (Angaut and Brodal 1967; Precht et al. 1976b). Nodular stimulation in the cat has a weak (much weaker than floccular stimulation) inhibitory effect on some vestibulo-ocular neurons and on vestibular nuclei neurons projecting to the cerebellum (Precht et al. 1976b).

The occurrence of spontaneous nystagmus and positional nystagmus have been observed with nodular lesions and stimulations (Aschan et al. 1964; Ron and Robinson 1973; Spiegel and Scala 1942). Habituation of vestibular nystagmus was almost absent after nodular lesions (pigeon, Halstead et al. 1937; cat, Singleton 1967; monkey, Waespe et al. 1985a). Stimulation of the nodulus inhibited ongoing caloric and vestibular nystagmus (Fernandez and Frederickson 1964). In contrast to floccular lesions, the strongest effect of nodular lesions is not on eye movements, but on postural control (Dow 1938).

Neuronal activity has not yet been recorded in the vermian part of the vestibulocerebellum in the monkey. In the rabbit PCs have been found to be responsive to natural vestibular and visual stimuli (Precht et al. 1976a).

6.3.1 Effects of Lesions on Eye Movements in the Monkey

After complete surgical nodulectomy and partial uvulectomy, OKN dynamics were not significantly changed (Waespe et al. 1985a). The initial fast increase in eye velocity as well as steady state velocities of OKN were similar before and after surgery. Igarashi et al. (1975), however, reported lower OKN steady state velocities at high stimulus velocities above $100^{\circ} \cdot s^{-1}$ postoperatively in the squirrel monkey. The initial velocity of OKAN was smaller after the operation and the time constant was significantly increased with values of up to 30–40 s (Waespe et al. 1985a). The time constant of OKAN and of VN could not be shortened by repetitive stimulation. Vestibular nystagmus was not changed in its gain, but its time constant was increased. The most dramatic effects of nodulectomy and uvulectomy were observed on visual and vestibular suppression of postrotatory nystagmus and OKAN. After the operation tilting of the animal had no effect on the time constant of postrotatory nystagmus and OKAN (Waespe et al. 1985a). Time constants during trials when the animal was tilted were the same as when the animal was upright (see Sect. 2.2.2). Exposure to the stationary visual surround during postrotatory nystagmus and OKAN caused the same rapid drop in eye velocity of nystagmus as before the lesion, but eye velocity did not decay rapidly to zero after the initial drop. When the lights were extinguished slow phase velocity recovered to values close to, but not identical with those which would have been expected had there been no period of suppression. Only after a period of visual suppression of 20–30 s (compared to 6–8 s before the operation) did the nystagmus not reappear when the animal was put into darkness. This suggests that the eye velocity storage mechanism can no longer be "dumped" after nodulectomy and uvulectomy.

Cohen and coworkers (Waespe et al. 1985a) suggested that the visually and vestibularly induced dynamic modification of the characteristics of the velocity storage is mediated through the nodulus and uvula of the vestibulocerebellum. Another finding was the occurrence of periodic alternating nystagmus in darkness after surgery. Spontaneous nystagmus developed either spontaneously or after stimulation. It changed direction every 3–4 min with peak velocities of $25-35^{o}\cdot s^{-1}$. In the rabbit ablation of the nodulus and uvula did not affect VOR dynamics, but the OKN gain was significantly increased (Nagao 1983).

6.3.2 Conclusion

The role of the flocculus during visual-vestibular interaction has been examined in much more detail than that of the nodulus. Lesion studies in the monkey suggest that the nodulus subserves quite different functions than the flocculus. The flocculus, nodulus, and uvula seem to have complementary roles in modifying the VOR. The flocculus mediates signals for rapid changes in eye velocity induced by the visual system. It aids the VOR in stabilizing gaze during ocular compensation and suppresses the VOR by fixation of a target or visual surround that moves with the head. The nodulus and uvula, on the other hand, appear to have little or no effect on visual and vestibular pathways that cause rapid changes in eye velocity. Instead they affect the VOR by dynamically controlling the characteristics of the velocity storage mechanism. In contrast to the flocculus, nothing is known about neural events in the nodulus and uvula.

6.4 Other Parts of the Cerebellum

In the monkey neuronal activity was recorded in the vermian lobule VI and VII. PC activity during smooth pursuit changed in phase with eye velocity as well as with image slip velocity of the target when the monkey fixated a stationary point. It was concluded that vermian PCs code target velocity, i.e., the sum of eye and retinal image slip velocity (Kase et al. 1979; Suzuki et al. 1981). A head velocity-related modulation of vermian PCs was also found (Suzuki and Keller 1982).

The vermian lobule VI and VII project to the nuclei fastigii (Haines 1977). Fastigial neurons are modulated by vestibular stimulation (Gardner and Fuchs 1975; Kimm et al. 1976), but their behavior during SP has not yet been tested.

7 General Conclusions

The vestibulo-ocular reflex (VOR) is a seemingly simple reflex involving two relay neurons in the vestibular and oculomotor nuclei. This reflex works so reliably that its absence in patients is considered as one of the criterions of brain death. For gaze stabilization under physiological conditions, the dynamics of the VOR are limited and VOR performance is improved and aided by visual and other motion cues.

Single cell recordings in alert animals show an extensive and complex interaction of multimodal motion information in the vestibular nuclei and the cerebellum. It is a challenging task to analyze and to understand the meaning of the multiple parallel information processing which is characteristic of the central nervous system. As it is very difficult to define the precise location of specific cells in the information flow by recording their activity, models are a necessary and useful analytical tool. This has been demonstrated most successfully by Robinson (1981b) in the oculomotor system. Models need a rigorous quantification of data. With models the possible logic of information flow in the process of sensorimotor transformation can be defined more easily. The identification of internal parameters in models is important for the interpretation of single unit recordings as most neurons, if they are not peripheral vestibular neurons or motoneurons, carry a complex signal which cannot be related to input or output parameters like head, eye, or image slip velocity only.

Another most useful tool has been the placing of lesions, either surgically or chemically. Studying the information processing after lesions with single unit recordings can determine the possible function of neuron populations in mediating eye movement command signals. That lesions can have complex and unexpected effects is shown by the fact that a vestibular nerve lesion leads to the abolishment of OKAN, although no visual signals are carried in the vestibular nerve. These experiments also establish a link to clinical neurology, where the study of deficits in patients leads the neurologist to the localization of central and peripheral lesions.

We had to confine our review to the discussion of gaze stabilization in the horizontal plane. In passing we mentioned some experiments which showed the interdependence of the horizontal with the vertical system and their mutal dependence on otolith input. To explore the vestibular-visual system in the three dimensions of space will be a new challenge.

8 Summary

1. The vestibulo-ocular reflex (VOR), the optokinetic nystagmus (OKN), and smooth pursuit (SP) eye movements supplement each other in gaze stabilization. The most important features of the horizontal VOR, and the changes in its dynamics by visual and other information, are summarized for primates (Sect. 2). Because rabbit and cat studies have also been extensive, data from these experiments are discussed, when relevant. The VOR is most effective in the high frequency range and the optokinetic system in the low frequency range of rotational stimuli. During low accelerations and during constant velocity rotation in light, visual information therefore complements the VOR. Visual information can be used to suppress the VOR when a visual target or field is moving together with the head. The dynamics of the VOR can be changed by habituation. During rotation about an off-vertical axis (OVAR) and during pitch-while-rotating, vestibular information about changes of the head in space originating in vertical canals and otoliths contribute to the horizontal VOR. In humans the gain of the VOR depends on the mental set of the subjects. Visual experience in early life is crucial for the proper development of the VOR.

2. Models have been helpful in identifying the conceptual problems and in interpreting eye movement responses and single cell recordings (Sect. 2.7). There is general agreement that OKN has two dynamically distinct components. Fast dynamics of OKN slow phase velocity are assumed to be generated by activation of "direct" visual-oculomotor pathways or of the smooth pursuit system. Slow dynamics are assumed to be generated by activation of "indirect" pathways which contain a storage element. The eye velocity storage element is responsible for the prolongation of the time constant of the VOR beyond that of activity changes in the vestibular nerve. It is also responsible for the occurrence of optokinetic after-nystagmus (OKAN).

The relationship of OKN and SP is not yet clear. SP eye movements are induced by movement of small objects and OKN by movement of a large visual field. SP can easily be performed in front of a stationary background. SP eye movements induce relative movement of this background which would under other conditions induce OKN. The decisive difference between OKN and SP seems not to be whether foveal or parafoveal regions of the retina are stimulated, but rather how visual information is processed centrally.

3. The results of single cell recordings and lesions in the pretectum, the vestibular nerve, and the vestibular nuclei as well as in the vestibulocerebellum are discussed. An understanding of information processing in these structures is necessary for the concept of visual-vestibular interaction and the hypothesis of two dynamically different visual-oculomotor pathways.

4. Although vestibular nerve activity is not modulated by visual information, sectioning of the nerve permanently abolishes OKAN and the slow dynamics of OKN (Sect. 3); this suggests that the eye velocity storage mechanism is thereby inactivated.

5. The accessory optic system (AOS) and other pretectal structures and pontine nuclei in the cat and monkey receive both a direct retinal input and an input from visual and visual-associative cortical regions (Sect. 4). Neurons in the nucleus of the optic tract (NOT) in monkeys are modulated not only by moving large fields but also by moving a single light dot if its image moves across the fovea. Lesion studies suggest that pretectal structures mediate visual information important for the generation of OKN and specifically of its slow dynamics. This information reaches the vestibular nuclei and the flocculus via pathways which are largely unknown.

6. There are several functionally distinct subgroups of neurons in the vestibular nuclear complex (Sect. 5). We have described activity changes in one of these subgroups and have termed the neurons "central vestibular neurons". These neurons are modulated by vestibular as well as by optokinetic stimulation, and the time course of their activity changes closely reflects that of VN and OKAN slow phase velocity. These "central vestibular neurons" have no or only a small eye position signal, and they are not or only minimally activated during SP eye movements. Activity changes of these neurons are interpreted to reflect the activity within the "indirect" visual-oculomotor pathways involving the eye velocity storage. It is assumed that these neurons are inactivated after a bilateral vestibular nerve cut (neurectomy). The activity of other neurons in the vestibular nuclei are strongly related to different eye positions. During SP, activity changes reflect eye position and eye position changes. These and other neurons may also be part of the velocity-to-position integrator.

One of the important problems is that the intrinsic connectivity of the neurons within the vestibular nuclei and with neurons in the cerebellum and brainstem is largely unknown. The precise functional role of these neurons in the sensorimotor transformation has therefore also not been defined. Lesion studies are of limited value as neurons with different physiological characteristics are intermingled. Furthermore, current terminology of classifying neurons according to physiological characteristics is rather unsatisfactory and is discussed.

7. The vestibulocerebellum has been assumed for a long time to be involved in the VOR and in the modification of its dynamics by visual and vestibular information and by habituating stimuli.

The functional roles (discussed in Sect. 6) of the flocculus and the nodulus seem to be, however, quite distinct from each other. The flocculus and the nodulus have dense reciprocal connections with the vestibular nuclei, but their target cells seem to be different. Lesion studies suggest that the flocculus is important in mediating signals for rapid changes of slow phase velocity of OKN and for SP. The notion that floccular PCs are involved in the "direct" visual-oculomotor pathway receives support from single cell recordings in normal monkeys and after vestibular neurectomy. Single cell recordings from the nodulus are not available. Based on interpretation of lesion studies, the nodulus is suggested to mediate signals which change the dynamics of the eye velocity storage mechanism.

Acknowledgments. We thank Mrs. V. Furrer-Isoviita for assistance during experiments, Mr. J. Müller for photographic work, and Mrs. B. Gähler for secretarial work.

References

Abend WK (1977) Functional organization of the superior vestibular nucleus of the squirrel monkey. Brain Res 132:65–84

Adler B, Collewijn H, Curio G, Grüsser O-J, Pause M, Schreiter U, Weiss L (1981) Sigma-movement and sigma-nystagmus: a new tool to investigate the gaze-pursuit system and visual-movement perception in man and monkey. Ann NY Acad Sci 374:284–302

Albus K, Donate-Oliver F, Sanides D, Fries W (1981) The distribution of pontine projection cells in visual and association cortex of the cat: an experimental study with horseradish peroxidase. J Comp Neurol 201:175–189

Allum JHJ, Graf W, Dichgans J, Schmidt CL (1976) Visual-vestibular interactions in the vestibular nuclei of the goldfish. Exp Brain Res 26:463–485

Angaut P, Brodal A (1967) The projection of the "vestibulocerebellum" onto the vestibular nuclei in the cat. Arch Ital Biol 105:441–479

Aschan G, Eksall L, Grant G (1964) Nystagmus following stimulation in the central vestibular pathways using permanent implanted electrodes. Acta Otolaryngol Suppl 192:63–77

Atkinson J (1979) Development of optokinetic nystagmus in the human infant and monkey infant: an analogue to development in kittens. In: Freeman RD (ed) Developmental Neurobiology of Vision. NATO Advanced Study Institute Series: Series A, Life Sciences. Plenum, New York, pp 277–287

Bach M, Bouis D, Fischer B (1983) An accurate and linear infrared oculometer. J Neurosci Methods 9:9–14

Baker J, Gibson A, Glickstein M, Stein J (1976) Visual cells in the pontine nuclei of the cat. J Physiol (Lond) 255:415–433

Baker R, Berthoz A (1975) Is the prepositus hypoglossi nucleus the source of another vestibular ocular pathway? Brain Res 86:121–127

Baker R, Berthoz A (eds) (1977) Control of gaze by brain stem neurons. Dev Neurosci vol 1. Elsevier, Amsterdam

Balaban CD (1983) A projection from nucleus reticularis tegmenti pontis of Bechterew to the medial vestibular nucleus in rabbits. Exp Brain Res 51:304–309

Balaban CD, Watanabe E (1984) Functional representation of eye movements in the flocculus of monkeys (*Macaca fuscata*). Neurosci Lett 49:199–205

Balaban CD, Ito M, Watanabe E (1981) Demonstration of zonal projections from the cerebellar flocculus to vestibular nuclei in monkeys (*Macaca fuscata*). Neurosci Lett 27:101–105

Baloh RW, Henn V, Jaeger J (1982) Habituation of the human vestibulo-ocular reflex by low frequency harmonic acceleration. Am J Otolaryngol 3:235–241

Barmack NH (1970) Dynamic visual acuity as an index of eye movement control. Vision Res 10:1377–1391

Barmack NH, Pettorossi VE (1985) Effects of unilateral lesions of the flocculus on optokinetic and vestibuloocular reflexes of the rabbit. J Neurophysiol 53: 481–496

Barnes GR, Benson AJ, Prior ARJ (1978) Visual-vestibular interaction in the control of eye movement. Aviat Space Environ Med 49:557–564

Barr CC, Schultheis LW, Robinson DA (1976) Voluntary, non-visual control of the human vestibulo-ocular reflex. Acta Otolaryngol 81:365–375

Behrens F, Grüsser O-J (1982) On the additivity of sigma- and phi-movement in visual perception and oculomotor control. Human Neurobiol 1:121–127

Bender MB, Shanzer S (1983) History of optokinetic nystagmus. Neuro Ophthalmol 3:73–88

Benevento LA, Rezak M, Santos-Anderson R (1977) An autoradiographic study of the projections of the pretectum in the rhesus monkey (*Macaca mulatta*): evidence for sensorimotor links to the thalamus and oculomotor nuclei. Brain Res 127:197–218

Berman N (1977) Connections of the pretectum in the cat. J Comp Neurol 174: 227–254

Berthoz A, Melvill-Jones (eds) (1985) Adaptive mechanisms in gaze control – facts and theories. Reviews of oculomotor research, vol 1. Elsevier, Amsterdam

Blanks RHI, Precht W (1983) Responses of units in the rat cerebellar flocculus during optokinetic and vestibular stimulation. Exp Brain Res 53:1–15

Blanks RHI, Estes MS, Markham C (1975) Physiologic characteristics of vestibular first-order canal neurons in the cat. II. Response to constant angular acceleration. J Neurophysiol 38:1250–1268

Blanks RHI, Volkind R, Precht W, Baker R (1977) Responses of cat prepositus hypoglossi neurons to horizontal angular acceleration. Neurosci 2:391–403

Blanks RHI, Precht W (1978) Response properties of vestibular afferents in alert cats during optokinetic and vestibular stimulation. Neurosci Lett 10:225–229

Blanks RHI, Precht W, Torigoe Y (1983) Afferent projections to the cerebellar flocculus in the pigmented rat demonstrated by retrograde transport of horseradish peroxidase. Exp Brain Res 52:293–306

Blanks RHI, Curthoys IS, Bennett ML, Markham CH (1985) Planar relationships of the semicircular canals in rhesus and squirrel monkeys. Brain Res 340:315–324
Böhmer A, Henn V (1983) Horizontal and vertical vestibulo-ocular and cervico-ocular reflexes in the monkey during high frequency rotation. Brain Res 277: 241–248
Böhmer A, Henn V, Suzuki J-I (1985) Vestibulo-ocular reflexes after selective plugging of the semicircular canals in the monkey. Response plane determinations. Brain Res 326:291–298
Bond HW, Ho P (1970) Solid miniature silver-silver chloride electrodes for chronic implantation. Electronencephalogr Clin Neurophysiol 28:206–208
Boyle R, Büttner U, Markert G (1985) Vestibular nuclei activity and eye movements in the alert monkey during sinusoidal optokinetic stimulation. Exp Brain Res 57:362–369
Brodal A (1954) Afferent cerebellar connections. In: Jansen J, Brodal A (eds) Aspects of cerebellar anatomy. JS Tanum, Oslo, pp 82–188
Brodal A (1972) Some features in the anatomical organization of the vestibular nuclear complex in the cat. In: Brodal A, Pompeiano O (eds) Basic aspects of central vestibular mechanisms, Prog Brain Res, vol 37. Elsevier, Amsterdam pp 31–53
Brodal A (1981) Neurological anatomy. Oxford University Press, Oxford
Brodal A (1983) The periphypoglossal nuclei in the macaque monkey and the chimpanzee. J Comp Neurol 218:257–269
Brodal A (1984) The vestibular nuclei in the macaque monkey. J Comp Neurol 227:252–266
Brodal A, Drahløs PA (1963) Two types of mossy fiber terminals in the cerebellum and their regional distribution. J Comp Neurol 121:173–187
Brodal A, Høivik B (1964) Site and mode of termination of primary vestibulo-cerebellar fibres in the cat. An experimental study with silver impregnation methods. Arch Ital Biol 102:1–21
Brodal A, Brodal P (1985) Observations on the secondary vestibulocerebellar projections in the macaque monkey. Exp Brain Res 58:62–79
Brodal P (1978) The corticopontine projection in the rhesus monkey: origin and principles of organisation. Brain 101:251–283
Brodal P (1982) Further observations on the cerebellar projections from the pontine nuclei and the nucleus reticularis pontis in the rhesus monkey. J Comp Neurol 204:44–55
Brodal P, Brodal A (1982) Further observations on the olivocerebellar projection in the monkey. Exp Brain Res 45:71–83
Buettner UW, Büttner U (1979) Vestibular nuclei activity in the alert monkey during suppression of vestibular and optokinetic nystagmus. Exp Brain Res 37:581–593
Buettner UW, Büttner U, Henn V (1978) Transfer characteristics of neurons in vestibular nuclei of the alert monkey. J Neurophysiol 41:1614–1628
Buettner UW, Henn V, Young LR (1981) Frequency response of the vestibulo-ocular reflex (VOR) in the monkey. Aviat Space Environment Med 52:73–77
Buizza A, Schmid R (1982) Visual-vestibular interaction in the control of eye movement: mathematical modeling and computer stimulation. Biol Cybern 43: 209–223
Bushnell MC, Goldberg ME, Robinson DL (1981) Behavioral enhancement of visual responses in monkey cerebral cortex. I. Modulation in posterior parietal cortex related to selective visual attention. J Neurophysiol 46:755–772
Büttner U, Waespe W (1981) Vestibular nerve activity in the alert monkey during vestibular and optokinetic nystagmus. Exp Brain Res 41:310–315

Büttner U, Waespe W (1984) Purkinje cell activity in the primate flocculus during optokinetic stimulation, smooth pursuit eye movements and VOR-suppression. Exp Brain Res 55:97–204

Büttner U, Meienberg D, Schimmelpfennig B (1983) The effect of central retinal lesions on optokinetic nystagmus in the monkey. Exp Brain Res 52:248–256

Cannon SC, Robinson DA (1986) The final common integrator is in the prepositus and vestibular nuclei. In: Keller EL, Zee DS (eds) Adaptive processes in visual and oculomotor systems. Pergamon Press pp 307–311

Carleton SC, Carpenter MB (1983) Afferent and efferent connections of the medial, inferior and lateral vestibular nuclei in the cat and monkey. Brain Res 278: 29–51

Carleton SC, Carpenter MB (1984) Distribution of primary vestibular fibers in the brainstem and cerebellum of the monkey. Brain Res 294:281–298

Carpenter MB, Stein BM, Peter P (1972) Primary vestibulocerebellar fibers in the monkey: distribution of fibers arising from distinctive cell groups of the vestibular ganglia. Am J Anat 135:221–250

Cazin L, Precht W, Lannou J (1982) Non-cerebellar visual afferents to the vestibular nuclei involving the prepositus hypoglossal complex: an autoradiographic study in the rat. Exp Brain Res 48:309–313

Cheron G, Gillis P, Godaux E (1986a) Lesions in the cat prepositus complex: effects on the optokinetic system. J Physiol 372:95–111

Cheron G, Godaux E, Laune JM, Vanderkelen B (1986b) Lesions in the cat prepositus complex: effects on the vestibulo-ocular reflex and saccades. J Physiol (Lond) 372:75–94

Chubb MC, Fuchs AF, Scudder CA (1984) Neuron activity in monkey vestibular nuclei during vertical vestibular stimulation and eye movements. J Neurophysiol 52:724–742

Cohen B (1974) The vestibulo-ocular reflex arc. In: Kornhuber HH (ed) Handbook of sensory physiology. Springer, Berlin Heidelberg New York, pp 478–540

Cohen B (ed) (1981) Vestibular and oculomotor physiology. Ann NY Acad Sci 374, New York

Cohen B, Uemura T, Takemori S (1973) Effects of labyrinthectomy on optokinetic nystagmus (OKN) and optokinetic after-nystagmus (OKAN). Equilibrium Res 3:88–93

Cohen B, Matsuo V, Raphan T (1977) Quantitative analysis of the velocity characteristics of optokinetic nystagmus and optokinetic after-nystagmus. J Physiol (Lond) 270:321–344

Cohen B, Henn V, Raphan T, Dennett D (1981) Velocity storage, nystagmus, and visual-vestibular interactions in humans. Ann NY Acad Sci 374:421–433

Cohen B, Suzuki J-I, Raphan T (1983) Role of the otolith organs in generation of horizontal nystagmus: effects of selective labyrinthine lesions. Brain Res 276: 159–164

Collewijn H (1975) Direction selective units in the rabbit's nucleus of the optic tract. Brain Res 100:489–508

Collewijn H (1977) Eye- and head movements in freely moving rabbits. J Physiol (Lond) 266:471–498

Collewijn H (1981) The oculomotor system of the rabbit and its plasticity. In: Braitenberg et al. (eds) Studies brain function. Springer, Berlin Heidelberg New York

Collewijn H, Tamminga EP (1984) Human smooth and saccadic eye movements during voluntary pursuit of different target motions on different backgrounds. J Physiol (Lond) 351:217–250

Collewijn H, Conijn P, Martins AJ, Tamminga EP, Van Die GC (1982) Control of gaze in man: synthesis of pursuit, optokinetic and vestibulo-ocular systems. In: Roucoux A, Crommelinck M (eds) Physiological and pathological aspects of eye movements. W Junk, The Hague, pp 3–22

Conway JL, Timberlake GT, Skavenski AA (1981) Oculomotor changes in cats reared without experiencing continuous retinal image motion. Exp Brain Res 43:229–232

Curthoys IS, Markham CH (1971) Convergence of labyrinth influences on units in the vestibular nuclei of the cat. I. Natural stimulation. Brain Res 35:469–490

Cynader M (1985) Effects of visual deprivation on properties and modifiability of compensatory eye movement systems. In: Berthoz A, Melvill Jones G (eds) Adaptive mechanisms in gaze control – facts and theories. Amsterdam, Elsevier, pp 95–109

Cynader M, Chernenko G (1976) Abolition of direction selectivity in the visual cortex of the cat. Science 193:504–505

Demer JL, Robinson DA (1983) Different time constants for optokinetic and vestibular nystagmus with a single velocity storage element. Brain Res 276:173–177

Dichgans J, Bizzi E, Morasso P, Tagliasco V (1973a) Mechanisms underlying recovery of eye-head coordination following bilateral labyrinthectomy in monkeys. Exp Brain Res 18:548–562

Dichgans J, Schmidt CL, Graf W (1973b) Visual input improves the speedometer function of the vestibular nuclei in the goldfish. Exp Brain Res 18:319–322

Dieringer N, Precht W (1982) Dynamics of compensatory vestibular reflexes in the grassfrog, *Rana temporaria*. In: Roucoux A, Crommelinck M (eds) Physiological and pathological aspects of eye movements. W Junk, The Hague, pp 417–423

Dodge R (1923) Adequacy of reflex compensatory eye movements including the effects of neural rivalry and competition. J Exp Psychol 6:169–181

Dow RS (1935) The relation of the paraflocculus to movements of the eyes. Am J Physiol 113:296–298

Dow RS (1936) The fiber connections of the posterior parts of the cerebellum in the rat and cat. J Comp Neurol 63:527–548

Dow RS (1938) Efferent connections of the flocculo-nodular lobe in *Macaca mulatta*. J Comp Neurol 68:279–305

Dubois MFW, Collewijn H (1979) Optokinetic reactions in man elicited by localized retinal motion stimuli. Vision Res 19:1105–1115

Duensing F, Schaefer K-P (1958) Die Aktivität einzelner Neurone im Bereich der Vestibulariskerne bei Horizontalbeschleunigungen unter besonderer Berücksichtigung des vestibulären Nystagmus. Arch Psychiat Nervenkr 198:225–252

Duensing F, Schaefer K-P (1960) Die Aktivität einzelner Neurone der Formatio reticularis des nicht gefesselten Kaninchens bei Kopfwendungen und vestibulären Reizen. Arch Psychiat Nervenkr 201:97–122

Eccles JC, Ito M, Szentagothai J (1967) The cerebellum as a neuronal machine. Springer, Berlin Heidelberg New York

Eckmiller R, Mackeben M (1980) Pre-motor single unit activity in the monkey brain stem correlated with eye velocity during pursuit. Brain Res 184:210–214

Epema AH, Guldemond JM, Voogd J (1985) Reciprocal connections between the caudal vermis and the vestibular nuclei in the rabbit. Neurosci Lett 57: 273–278

Estanol B, Romero R, Corvery J (1979) Effects of cerebellectomy on eye movements in main. Arch Neurol 36:281–284

Ewald JR (1892) Physiologische Untersuchungen über das Endorgan des Nervus octavus. Bergmann, Wiesbaden

Fernandez C, Frederickson JM (1964) Experimental cerebellar lesions and their effect on vestibular function. Acta Otolaryngol Suppl 192:52–62

Fernandez C, Goldberg JM (1971) Physiology of peripheral neurons innervating semicircular canals of the squirrel monkey. II. Response to sinusoidal stimulation and dynamics of peripheral vestibular system. J Neurophysiol 34:661–675

Flandrin JM, Courjon JH, Jeannerod M, Schmid R (1983) Effects of unilateral flocculus lesions on vestibulo-ocular responses in the cat. Neurosci 8:809–817

Fuchs AF (1967) Saccadic and smooth pursuit eye movements in the monkey. J Physiol (Lond) 191:609–631
Fuchs AF, Becker W (eds) (1981) Progress in oculomotor research. Dev Neurosci 12. Elsevier, Amsterdam
Fuchs AF, Kimm P (1975) Unit activity in vestibular nuclei of the alert monkey during horizontal angular acceleration and eye movement. J Neurophysiol 38:1140–1161
Furman JM, O'Leary DP, Wolfe JW (1982) Dynamic range of the frequency response of the horizontal vestibulo-ocular reflex of the alert rhesus monkey. Acta Otolaryngol 93:81–91
Gacek RR (1969) The course and central termination of first oder neurons supplying vestibular end organs in the cat. Acta Otolaryngol Suppl 254:1–66
Gacek RR, Lyon M (1974) Localization of vestibular efferent neurons in the kitten with horseradish perxoidase. Acta Otolaryngol 77:92–101
Gacek RR, Rasmussen GL (1961) Fiber analysis of the stato-acoustic nerve of the guinea pig, cat, and monkey. Anat Rec 139:445–463
Galiana HL (1986) A new approach to understanding adaptive visual-vestibular interactions in the central nervous system. J Neurophysiol 55:349–377
Gardner EP, Fuchs AF (1975) Single-unit responses to natural vestibular stimuli and eye movements in deep cerebellar nuclei of the alert rhesus monkey. J Neurophysiol 38:627–649
Gerrits NM, Epema AH, Voogd J (1984) The mossy fiber projection of the nucleus reticularis tegmenti pontis to the flocculus and adjacent ventral paraflocculus in the cat. Neurosci 11:627–644
Giolli RA (1963) An experimental study of the accessory optic system in the cynomolgus monkey. J Comp Neurol 121:89–107
Glickstein M, May JG III (1982) Visual control of movement: the circuits which link visual input to the pons and cerebellum. In: Neff WD (ed) Contributions to sensory physiology. Academic Press, New York, pp 103–145
Glickstein M, Stein J, King RA (1972) Visual input to the pontine nuclei. Science 178:1110–1111
Glickstein M, Cohen JL, Dixon B, Gibson A, Hollins M, Labossiere E, Robinson F (1980) Corticopontine visual projections in macaque monkeys. J Comp Neurol 190:209–229
Godaux E, Vanderkelen B (1984) Vestibulo-ocular reflex, optokinetic response and their interactions in the cerebellectomized cat. J Physiol (Lond) 346:155–170
Goldberg JM, Fernandez C (1971a) Physiology of peripheral neurons innervating semicircular canals of the squirrel monkey. I. Resting discharge and response to constant angular accelerations. J Neurophysiol 34:635–660
Goldberg JM, Fernandez C (1971b) Physiology of peripheral neurons innervating semicircular canals of the squirrel monkey. III. Variations among units in their discharge properties. J Neurophysiol 34:676–684
Goldberg JM, Fernandez C (1975) Vestibular mechanisms. Ann Rev Physiol 37: 129–162
Goldberg JM, Fernandez C (1980) Efferent vestibular system in the squirrel monkey: anatomical location and influence on afferent activity. J Neurophysiol 43: 986–1025
Goldberg JM, Fernandez C (1982) Eye movements and vestibular-nerve responses produced in the squirrel monkey by rotations about an earth-horizontal axis. Exp Brain Res 46:393–402
Grasse KL, Cynader MS (1984) Electrophysiology of lateral and dorsal terminal nuclei of the cat accessory optic system. J Neurophysiol 51:276–293
Grüsser O-J, Pause M, Schreiter U (1979) Three methods to elicit Sigma-optokinetic nystagmus in Java monkeys. Exp Brain Res 35:519–526
Guedry FE (1965) Orientation of the rotation axis relative to gravity: its influence on nystagmus and the sensation of rotation. Acta Otolaryngol 60:30–48

Haines DE (1977) Cerebellar corticonuclear and corticovestibular fibers of the flocculonodular lobe in a prosimian primate (*Galago senegalensis*). J Comp Neurol 174:607–630
Halstead W, Yacorzynski G, Fearing F (1937) Further evidence of cerebellar influence in the habituation of after-nystagmus in pigeons. Am J Physiol 120:350–355
Harris LR, Lepore F, Guillemot JP, Cynader M (1980) Abolition of optokinetic nystagmus in the cat. Science 210: 91–92
Henn V, Young LR, Finley C (1974) Vestibular nucleus units in alert monkeys are also influenced by moving visual fields. Brain Res 71:144–149
Henn V, Cohen B, Young LR (1980) Visual-vestibular interaction in motion perception and the generation of nystagmus. Neurosci Res Prog Bull 18:457–651
Henn V, Reisine H, Waespe W, Boehmer A (1983) Pathophysiology of the vestibular system and its clinical implication. In: Suzuki J-I (ed) Clinical examination on vertigo and loss-of-balance. Gendai, Tokyo, pp 366–378
Henn V, Lang W, Hepp K, Reisine H (1984) Experimental gaze palsies in monkeys and their relation to human pathology. Brain 107:619–636
Hoddevik GH (1978) The projection from nucleus reticularis tegmenti pontis onto the cerebellum in the cat. A study using the methods of anterograde degeneration and retrograde axonal transport of horseradish peroxidase. Anat Embryol 153:227–242
Hoffmann K-P (1982) Cortical versus subcortical contributions to the optokinetic reflex in the cat. In: Lennerstrand G (ed) Functional basis of ocular motility disorders. Pergamon Press, Oxford, pp 303–310
Hoffmann K-P, Distler C (1986) The role of direction selective cells in the nucleus of the optic tract of the cat and monkey during optokinetic nystagmus. In: Keller EL, Zee DS (eds) Adaptive processes in visual and oculomotor system. Pergamon Press Oxford, pp 261–266
Hoffmann K-P, Schoppmann A (1975) Retinal input to direction selective cells in the nucleus tractus opticus of the cat. Brain Res 99: 359–366
Honrubia V, Koehn WW, Jenkins HA, Fenton WH (1982) Effect of bilateral ablation of the vestibular cerebellum on visual-vestibular interaction. Exp Neurol 75: 616–626
Hudspeth AJ (1983) The hair cells of the inner ear. Sci American 248 1:42–52
Hutchins B, Weber JT (1985) The pretectal complex of the monkey: a reinvestigation of the morphology and retinal terminations. J Comp Neurol 232:425–442
Igarashi M, Miyata H, Kato Y, Wright WK, Levy JK (1975) Optokinetic nystagmus after cerebellar uvulanodulectomy in squirrel monkeys. Acta Otolaryngol 80: 180–184
Igarashi M, Takahashi M, Kubo T, Alford BR, Wright WK (1980) Effect of off-vertical tilt and macular ablation on postrotatory nystagmus in the squirrel monkey. Acta Otolaryngol 90:93–99
Ingvar S (1918) Zur Phylo- und Ontogenese des Kleinhirns nebst einem Versuche zu einheitlicher Erklärung der zerebellären Funktion und Lokalisation. Folia Neuro-Biol (Leipzig) 11:205–495
Ito M (1970) Neurophysiological aspects of the cerebellar motor control system. Int J Neurol 7:162–176
Ito M (1984) The cerebellum and neural control. Raven, New York
Ito M, Jastreboff PJ, Miyashita Y (1982) Specific effects of unilateral lesions in the flocculus upon eye movements of rabbits. Exp Brain Res 45:233–242
Jaeger J, Henn V (1981) Habituation of the vestibulo-ocular reflex (VOR) in the monkey during sinusoidal rotation in the dark. Exp Brain Res 41:108–114
Jaeger J, Henn V, Lang W, Miles TS, Waespe W (1981) Vestibular unit activity in monkeys with horizontal gaze palsy. In: Fuchs AF, Becker W (eds) Progr Oculomot Res, Dev Neurosci, vol 12. Elsevier, Amsterdam, pp 89–95
Janeke JB, Jongkees LBW, Oosterveld WJ (1970) Relationship between otoliths and nystagmus. Acta Otolaryngol 69:1–6

Jensen DW (1983) Survival of function in the deafferented vestibular nerve. Brain Res 273:175–178
Jung R (1948) Die Registrierung des postrotatorischen und optokinetischen Nystagmus und die optisch-vestibuläre Integration beim Menschen. Acta Otolaryngol 36:199–202
Jung R (1978) Perception, consciousness and visual attention. In: Buser P, Rougeul-Buser A (eds) Cerebral correlates of conscious experience. INSERM Symp 6. Elsevier, Amsterdam, pp 15–36
Kasai T, Zee DS (1978) Eye-head coordination in labyrinthine-defective human beings. Brain Res 144:123–141
Kase M, Noda H, Suzuki DA, Miller DC (1979) Target velocity signals of visual tracking in vermal Purkinje cells of the monkey. Science 205:717–720
Kato I, Harada K, Nakamura T, Sato Y, Kawasaki T (1982) Role of the nucleus reticularis pontis on visually induced eye movements. Exp Neurol 78:503–516
Kato I, Harada K, Hasegawa T, Igarashi T, Koike Y, Kawasaki T (1986) Role of the nucleus of the optic tract in monkeys in relation to optokinetic nystagmus. Brain Res 364:12–22
Keller EL (1976) Behavior of horizontal semicircular canal afferents in alert monkey during vestibular and optokinetic stimulation. Exp Brain Res 24:459–471
Keller EL (1978) Gain of the vestibulo-ocular reflex in monkey at high rotational frequencies. Vision Res 18:311–315
Keller EL, Crandall WF (1983) Neuronal responses to optokinetic stimuli in pontine nuclei of behaving monkey. J Neurophysiol 49:169–187
Keller EL, Daniels PD (1975) Oculomotor related interaction of vestibular and visual stimulation in vestibular nucleus cells in alert monkey. Exp Neurol 46: 187–198
Keller EL, Kamath BY (1975) Characteristics of head rotation and eye movement-related neurons in alert monkey vestibular nucleus. Brain Res 100:182–187
Keller EL, Precht W (1979) Visual-vestibular responses in vestibular nuclear neurons in the intact and cerebellectomized, alert cat. Neurosci 4:1599–1613
Kimm J, Hassul M, Cogdell B (1976) Fastigial neuronal responses to sinusoidal horizontal rotation. Exp Neurol 50:579–594
Koehn WW, Jenkins HA, Honrubia V, Fenton WH (1981) Effect of unilateral ablation of the vestibular cerebellum on visual-vestibular interaction. Exp Brain Res 73:618–631
Koenig E, Dichgans J (1981) Aftereffects of vestibular and optokinetic stimulation and their interaction. In: Cohen B (ed) Vestibular and oculomotor physiology. Ann NY Acad Sci 374:434–445
Koerner F, Schiller PH (1972) The optokinetic response under open and closed loop conditions in the monkey. Exp Brain Res 14:318–330
Kommerell G (ed) (1978) Augenbewegungsstörungen, Neurophysiologie und Klinik – Disorders of ocular motility, Neurophysiological and clinical aspects. Bergmann, München
Kommerell G, Täumer R (1972) Investigations of the eye tracking system through stabilised retinal images. Bibl Ophthalmol 82:288–297
Korte GE (1979) The brainstem projection of the vestibular nerve in the cat. J Comp Neurol 184:279–292
Korte GE, Friedrich VL (1979) The fine structure of the feline superior vestibular nucleus: identification and synaptology of the primary vestibular afferents. Brain Res 176:3–32
Korte GE, Mugnaini E (1979) The cerebellar projection of the vestibular nerve in the cat. J Comp Neurol 184:265–278
Korte GE, Friedrich VL (1979) The fine structure of the feline superior vestibular nucleus: identification and synaptology of the primary vestibular afferents. Brain Res 176:3–32

Kotchabhakdi N, Walberg F (1978) Primary vestibular afferent projections to the cerebellum as demonstrated by retrograde axonal transport of horseradish peroxidase. Brain Res 142:142–146
Kowler E, van den Steen J, Tamminga EP, Collewijn H (1984) Voluntary selection of the target for smooth eye movements in the presence of superimposed, full-field stationary, and moving stimuli. Vison Res 24:1789–1798
Lackner JR, Graybiel A (1981) Variations in gravitoinertial force level affect the gain of the vestibulo-ocular reflex: implications for the etiology of space motion sickness. Aviat Space Environ Med 52:154–158
Langer TP (1985) Basal interstitial nucleus of the cerebellum: cerebellar nucleus related to the flocculus. J Comp Neurol 235:38–47
Langer TP Fuchs AF, Chubb MC, Scudder CA, Lisberger SG (1985a) Floccular efferents in the rhesus macaque as revealed by autoradiography and horseradish peroxidase. J Comp Neurol 235:26–37
Langer TP, Fuchs AF, Scudder CA, Chubb MC (1985b) Afferents to the flocculus of the cerebellum in the rhesus macaque as revealed by retrograde transport of horseradish peroxidase. J Comp Neurol 235:1–25
Lannou J, Cazin L, Precht W, Letaillanter M (1984) Responses of prepositus hypoglossi neurons to optokinetic and vestibular stimulation in the rat. Brain Res 301:39–45
Larsell O (1937) The cerebellum. A review and interpretation. Arch Neurol Psychiat 38:580–607
Lennerstrand G, Zee DS, Keller EL (eds) (1982) Functional basis of ocular motility disorders. Wenner-Gren Symp Ser 37. Pergamon, Oxford
Lin H, Giolli RA (1979) Accessory optic system of rhesus monkey. Exp Neurol 63:163–176
Lindeman HH (1970) Studies on the morphology of the sensory regions of the vestibular apparatus. Ergeb Anat Entw Gesch 42:1–113
Lisberger SG, Fuchs AF (1978a) Role of primate flocculus during rapid behavioral modification of vestibuloocular reflex. I. Purkinje cell activity during visually guided horizontal smooth-pursuit eye movements and passive head rotation. J Neurophysiol 41:733–763
Lisberger SG, Fuchs AF (1978b) Role of primate flocculus during rapid behavioral modification of vestibuloocular reflex. II. Mossy fiber firing pattern during horizontal head rotation and eye movement. J Neurophysiol 41:764–777
Lisberger SG, Miles FA (1980) Role of primate medial vestibular nucleus in long-term adaptive plasticity of vestibuloocular reflex. J Neurophysiol 43:1725–1745
Lisberger SG, Pavelko TA (1986) Vestibular signals carried by the pathways subserving plasticity of the vestibulo-occular reflex in monkeys. Neurosci 6:346–354
Lisberger SG, Westbrook LE (1985) Properties of visual inputs that initiate horizontal smooth pursuit eye movements in monkeys. J Neurosci 5:1662–1673
Lisberger SG, Miles FA, Optican LM, Eighmy BB (1981) Optokinetic response in monkey: underlying mechanisms and their sensitivity to long-term adaptive changes in vestibuloocular reflex. J Neurophysiol 45:869–890
Lopez-Barneo J, Darlot C, Berthoz A (1981) Functional role of the prepositus hypoglossi nucleus in the control of gaze. In: Granit R, Pompeiano O (eds) Reflex control of posture and movement. Prog Brain Res Vol 50. Elsevier, Amsterdam; pp 667–679
Lopez-Barneo J, Darlot C, Berthoz A, Baker R (1982) Neuronal activity in prepositus nucleus correlated with eye movement in the alert cat. J Neurophysiol 47:329–352
Lorente de Nó R (1931) Ausgewählte Kapitel aus der vergleichenden Physiologie des Labyrinths. Die Augenmuskelreflexe beim Kaninchen und ihre Grundlagen. Ergebn Physiol 32:73–242

Lorente de Nó R (1933) Vestibulo-ocular reflex arc. Arch Neurol Psychiatr 30: 245–291

Lorente de Nó R (1938) Analysis of the activity of the chains of internuncial neurons. J Neurophysiol 1:207–244

Lorente de Nó R (1939) Transmission of impulses through cranial motor nuclei. J Neurophysiol 2:402–464

Louie AW, Kimm J (1976) The response of 8th nerve fibers to horizontal sinusoidal oscillation in the alert monkey. Exp Brain Res 24:447–457

Lynch JC, Mountcastle VB, Talbot WH, Yin TCT (1977) Parietal lobe mechanisms for directed visual attention. J Neurophysiol 40:362–389

Mach E (1875) Grundlinien der Lehre von den Bewegungsempfindungen. Engelmann, Leipzig

Madigan JC, Carpenter MB (1971) Cerebellum of the rhesus monkey. Univ Park Press, Baltimore

Maekawa K, Kimura M, Takeda T (1981a) Mossy fiber activation of the cerebellar flocculus from the visual system. Ann NY Acad Sci 374:476–490

Maekawa K, Takeda T, Kimura M (1981b) Neural activity of nucleus reticularis tegmenti pontis – the origin of visual mossy fiber afferents to the cerebellar flocculus of rabbits. Brain Res 210:17–30

Maekawa K, Takeda T, Kimura M (1984) Responses of the nucleus of the optic tract neurons projecting to the nucleus reticularis tegmenti pontis upon optokinetic stimulation in the rabbit. Neurosci Res 2:1–25

Magnin M, Courjon JH, Flandrin JM (1983) Possible visual pathways to the cat vestibular nuclei involving the nucleus prepositus hypoglossi. Exp Brain Res 51:298–303

Marcotte RR, Updyke BV (1982) Cortical visual areas of the cat project differentially onto the nuclei of the accessory optic system. Brain Res 242:205–217

McCrea RA, Baker R (1985a) Cytology and intrinsic organization of the perihypoglossal nuclei in the cat. J Comp Neurol 237:360–376

McCrea RA, Baker R (1985b) Anatomical connections of the nucleus prepositus of the cat. J Comp Neurol 237:377–407

McCrea RA, Baker R, Delgado-Garcia J (1979) Afferent and efferent organization of the prepositus hypoglossi nucleus. In: Granit R, Pompeiano O (eds) Reflex control of posture and movements. Prog Brain Res, Vol 50. Elsevier, Amsterdam; pp 653–665

McKinley PA, Peterson BW (1985) Voluntary modulation of the vestibuloocular reflex in humans and its relation to smooth pursuit. Exp Brain Res 60:454–464

Melvill Jones G, Milsum JH (1971) Frequency response analysis of central vestibular unit activity resulting from rotational stimulation of the semicircular canals. J Physiol (Lond) 219:191–215

Meyer CH, Lasker AG, Robinson DA (1985) The upper limit of human smooth pursuit velocity. Vision Res 25:561–563

Miles FA (1974) Single unit firing patterns in the vestibular nuclei related to voluntary eye movements and passive body rotation in conscious monkeys. Brain Res 71:215–224

Miles FA, Braitman DJ (1980) Long-term adaptive changes in primate vestibuloocular reflex. II. Electrophysiological observations on semicircular canal primary afferents. J Neurophysiol 43:1426–1436

Miles FA, Eighmy BB (1980) Long-term adaptive changes in primate vestibuloocular reflex. I. Behavioral observations. J Neurophysiol 43:1406–1425

Miles FA, Braitman DJ, Dow BM (1980a) Long-term adaptive changes in primate vestibuloocular reflex. IV. Electrophysiological observations in flocculus of adapted monkeys. J Neurophysiol 43:1477–1493

Miles FA, Fuller JH, Braitman DJ, Dow BM (1980b) Long-term adaptive changes in primate vestibuloocular reflex. III. Electrophysiological observations in flocculus of normal monkeys. J Neurophysiol 43:1437–1476

Miles FA, Kawano K, Optican LM (1986) Short latency ocular following responses of monkey. I. Dependence on temporo-spatial properties of the visual input. J Neurophysiol 56:1321–1354
Mitsacos A, Reisine H, Highstein SM (1983a) The superior vestibular nucleus: an intracellular HRP study in the cat. I. Vestibulo-ocular neurons. J Comp Neurol 215:78–91
Mitsacos A, Reisine H, Highstein SM (1983b) The superior vestibular nucleus: an intracellular HRP study in the cat. II. Non-vestibulo-ocular neurons. J Comp Neurol 215:92–107
Miyashita Y (1984) Eye velocity responsiveness and its proprioceptive component in the floccular Purkinje cells of the alert pigmented rabbit. Exp Brain Res 55:81–90
Miyashita Y, Nagao S (1984) Analysis of signal content of Purkinje cell responses to optokinetic stimuli in the rabbit cerebellar flocculus by selective lesions of brainstem pathways. Neurosci Res 1:223–241
Miyashita Y, Ito M, Jastreboff PJ, Maekawa K, Nagao S (1980) Effect upon eye movements of rabbits induced by severance of mossy fiber visual pathway to the cerebellar flocculus. Brain Res 198:210–215
Money KE, Scott JW (1962) Functions of separate sensory receptors of non-auditory labyrinth of the cat. Am J Physiol 202:1211–1220
Mowrer OH (1937) The influence of vision during bodily rotation upon the duration of post-rotational vestibular nystagmus. Acta Otolaryngol 25:351–364
Muratore R, Zee DS (1979) Pursuit after-nystagmus. Vision Res 19: 1057–1059
Naegele JR, Held R (1982) The postnatal development of monocular optokinetic nystagmus in infants. Vision Res 22:391–397
Nagao S (1983) Effects of vestibulocerebellar lesions upon dynamic characteristics and adaptation of vestibuloocular and optokinetic responses in pigmented rabbits. Exp Brain Res 53:36–46
Noda H (1981) Visual mossy fiber inputs to the flocculus of the monkey. In: Cohen B (ed) Vestibular and oculomotor physiology. Ann NY Acad Sci, vol 374. New York Acad Sci, New York; pp 465–475
Noda H, Suzuki DA (1979a) The role of the flocculus of the monkey in fixation and smooth pursuit eye movements. J Physiol (Lond) 294:335–348
Noda H, Suzuki DA (1979b) Processing of eye movement signals in the flocculus of the monkey. J Physiol (Ldon) 294:349–364
Noda H, Warabi T (1982) Eye position signals in the flocculus of the monkey during smooth-pursuit eye movements. J Physiol (Lond) 324:187–202
Ohm J (1933) Über die Beziehungen zwischen willkürlichen, optischen und vestibulären Augenbewegungen. Z Hals-Nasen-Ohrenheilk 32:234–246
Paige GD (1983) Vestibulo-ocular reflex and its interactions with visual following mechanisms in the squirrel monkey. I. Response characteristics in normal animals. J Neurophysiol 49:134–151
Pola J, Wyatt HJ (1980) Target position and velocity: the stimuli for smooth pursuit eye movements. Vision Res 20:523–534
Precht W (1978) Neuronal operations in the vestibular system. In: Braitenberg V (ed) Studies of brain function, vol 2. Springer, Berlin Heidelberg New York
Precht W, Simpson JI, Llinas R (1976a) Responses of Purkinje cells in rabbit nodulus and uvula to natural vestibular and visual stimuli. Pflügers Arch 367:1–6
Precht W, Volkind R, Maeda M, Giretti ML (1976b) The effects of stimulating the cerebellar nodulus in the cat on the responses of vestibular neurons. Neurosci 1:301–312
Precht W, Cazin L, Blanks R, Lannou J (1982) Anatomy and physiology of the optokinetic pathways to the vestibular nuclei in the rat. In: Roucoux A, Crommelinck M (eds) Physiological and pathological aspects of eye movements. W Junk, The Hague, pp 153–172

Precht W, Blanks RHI, Strata P, Montarolo P (1985) On the role of subprimate cerebellar flocculus in the optokinetic reflex and visual-vestibular interaction. In: Cerebellar functions. Bloedel JR, Dichgans J, Precht W (eds) Springer, Berlin, New York, pp 86–108

Raphan T, Cohen B (1985) Velocity storage and the ocular response to multidimensional vestibular stimuli. In: Berthoz A, Melvill Jones G (eds) Adaptive mechanisms in gaze control. Elsevier, Amsterdam; pp 123–143

Raphan T, Cohen B, Matsuo V (1977) A velocity-storage mechanism responsible for optokinetic nystagmus (OKN), optokinetic afternystagmus (OKAN) and vestibular nystagmus. In: Baker R, Berthoz A (eds) Control of gaze by brain stem neurons. Development Neurosci, vol I. Elsevier, Amsterdam; pp 37–47

Raphan T, Matsuo V, Cohen B (1979) Velocity storage in the vestibulo-ocular reflex arc (VOR). Exp Brain Res 35:229–248

Raphan T, Cohen B, Henn V (1981) Effects of gravity on rotatory nystagmus in monkeys. In: Cohen B (ed) Vestibular and oculomotor physiology. Ann NY Acad Sci, vol 374. New York Acad Sci, New York; pp 44–55

Raphan T, Cohen B, Suzuki J-I, Henn V (1983a) Nystagmus generated by pitch while rotating. Brain Res 276:165–172

Raphan T, Waespe W, Cohen B (1983b) Labyrinthine activation during rotation about axes tilted from the vertical. Adv Otorhinolaryngol 30:50–53

Rashbass C (1961) The relationship between saccadic and smooth tracking eye movements. J Physiol (Lond) 159:326

Reisine H, Simpson JI, Rudinger D, Henn V (1985) Combined anatomical and physiological study of semicircular canal orientation in the rhesus monkey. Soc Neurosci Abstr 11:319

Robinson DA (1963) A method of measuring eye movement using a scleral search coil in a magnetic field. IEEE Trans Biomed Electron 10:137–145

Robinson DA (1965) The mechanism of human smooth pursuit eye movement. J Physiol (Lond) 180:569

Robinson DA (1977) Linear addition of optokinetic and vestibular signals in the vestibular nucleus. Exp Brain Res 30:447–450

Robinson DA (1981a) The use of control systems analysis in the neurophysiology of eye movements. Ann Rev Neurosci 4:463–503

Robinson DA (1981b) Control of eye movements. In: VB Brooks (ed) Handbook of Physiology, the Nervous System, vol II, part 2. Williams & Wilkins, Baltimore, 1275–1320

Robinson DA (1982) A model of cancellation of the vestibulo-ocular reflex. In: Lennerstrand G, Zee DS, Keller EL (eds) Functional basis of ocular motility disorders. Pergamon, Oxford; pp 5–13

Ron S, Robinson DA (1973) Eye movements evoked by cerebellar stimulation in alert monkey. J Neurophysiol 36:1004–1022

Roucoux A, Crommelinck M (eds) (1982) Physiological and pathological aspects of eye movements. W Junk, The Hague

Rubertone JA, Haines DE (1981) Secondary vestibulocerebellar projections to flocculonodular lobe in a prosimian primate, *Galago senegalensis.* J Comp Neurol 200:255–272

Sandeman D (1983) The balance and visual systems of the swimming crab: their morphology and interaction. In: Horn E (ed) Fortschritte der Zoologie: Multimodal convergences in sensory systems. Gustav Fischer, Stuttgart; pp 213–229

Sato Y, Kawasaki T, Ikarashi K (1982) Zonal organization of the floccular Purkinje cells projecting to the vestibular nucleus in cats. Brain Res 232:1–15

Sato Y, Kawasaki T, Ikarashi K (1983a) Afferent projections from the brainstem to the three floccular zones in cats. I. Climbing fiber projections. Brain Res 272:27–36

Sato Y, Kawasaki T, Ikarashi K (1983b) Afferent projections from the brainstem to the three floccular zones in cats. II. Mossy fiber projections. Brain Res 272: 37–48
Scalia F (1972) The termination of retinal axons in the pretectal region of mammals. J Comp Neurol 145:223–258
Schaefer K-P, Zierau H, Suss KJ (1977) Differentiation of neuronal activity in the vestibular nuclei of rabbits. In: Baker R, Berthoz A (eds) Control of gaze by brain stem neurons. Dev Neurosci I. Amsterdam, Elsevier; pp 257–260
Schneider LW, Anderson DJ (1976) Transfer characteristics of first and second order lateral canal vestibular neurons in gerbil. Brain Res 112:61–76
Schrader V, Koenig E, Dichgans J (1985) The effect of lateral head tilt on horizontal postrotatory nystagmus I and II and the Purkinje effect. Acta Otolaryngol 100:98–105
Schuknecht HF (1982) Behavior of the vestibular nerve following labyrinthectomy. Ann Otol Rhino Laryngol (Suppl) 91:16–32
Segal BN, Liben S (1985) Modulation of human velocity storage sampled during intermittently-illuminated optokinetic stimulation. Exp Brain Res 59:515–523
Shimazu H, Precht W (1965) Tonic and kinetic responses of cat's vestibular neurons to horizontal angular acceleration. J Neurophysiol 28:991–1013
Shimazu H, Precht W (1966) Inhibition of central vestibular neurons from the contralateral labyrinth and its mediating pathway. J Neurophysiol 29:989–1013
Shinoda Y Yoshida K (1974) Dynamic characteristics of responses to horizontal head angular acceleration in vestibuloocular pathway in the cat. J Neurophysiol 37:653–673
Simpson JI (1984) The accessory optic system. Ann Rev Neurosci 7:13–41
Simpson JI, Graf W (1985) The selection of reference frames by nature and its investigators. In: Berthoz A, Melvill Jones G (eds) Adaptive mechanisms in gaze control – facts and theories. Rev Oculomotor Res, vol 1. Elsevier, Amsterdam; pp 3–16
Singleton GT (1967) Relationships of the cerebellar nodulus to vestibular function: a study of the effects of nodulectomy on habituation. Laryngoscope 77:1579–1620
Sirkin DW, Precht W, Courjon JH (1984) Initial, rapid phase of recovery from unilateral vestibular lesion in rat not dependent on survival of central portion of vestibular nerve. Brain Res 302:245–256
Skavenski AA, Robinson DA (1973) Role of abducens neurons in vestibuloocular reflex. J Neurophysiol 36:724–738
Smith E (1903) On the morphology of the brain in the mammalia, with special reference to that of the Lemurs, recent and extinct. Trans Linnean Soc 8:319
Spiegel EA, Scala NP (1942) Position nystagmus in cerebellar lesions. J Neurophysiol 5:247–260
Steiger J-J, Büttner-Ennever JA (1979) Oculomotor nucleus afferents in the monkey demonstrated with hoseradish peroxidase. Brain Res 160:1–15
Steinbach MJ (1976) Pursuing the perceptual rather than the retinal stimulus. Vision Res 16:1371
Steinman RM, Cushman WB, Martins AJ (1982) The precision of gaze. Human Neurobiol 1:97–109
Stroud BB (1895) The mammalian cerebellum. J Comp Neurol 5:71–118
Suzuki DA, Keller DA (1982) Vestibular signals in the posterior vermis of the alert monkey cerebellum. Exp Brain Res 47:145–147
Suzuki DA, Keller EL (1984) Visual signals in the dorsolateral pontine nucleus of the alert monkey: their relationship to smooth-pursuit eye movements. Exp Brain Res 53:473–478
Suzuki DA, Noda H, Kase M (1981) Visual and pursuit eve movement-related activity in posterior vermis of monkey cerebellum. 46:1120–1139

Takemori S, Cohen B (1974) Loss of visual suppression of vestibular nystagmus after flocculus lesions. Brain Res 72:213–224
Tauber ES, Atkin A (1968) Optomotor responses to monocular stimulation: relation to visual system organization. Science 160:1365–1367
Ter Braak JWG (1936) Untersuchungen über optokinetischen Nystagmus (translated "Investigations on optokinetic nystagmus" In: Collewijn H (1981) The oculomotor system of the rabbit and its plasticity. Studies of brain function, vol 5. Springer, Berlin Heidelberg New York) Arch Neerl Physiol 21:309–376
Tomlinson RD, Robinson DA (1984) Signals in vestibular nucleus mediating vertical eye movements in the monkey. J Neurophysiol 51:1121–1136
Uemura T, Cohen B (1973) Effects of vestibular nuclei lesions on vestibulo-ocular reflexes and posture in monkeys. Acta Otolaryngol (Suppl) (Stockh) 315:1–71
Van Die G, Collewijn H (1982) Optokinetic nystagmus in man. Human Neurobiol 1:111–119
Viirre E, Tweed D, Milner K, Vilis T (1986) A re-examination of the gain of the vestibulo-ocular reflex. J Neurophysiol 56:439–450
Waespe W, Cohen B (1983) Effects of flocculectomy on unit activity in the vestibular nuclei during visual-vestibular interactions. Exp Brain Res 51:23–35
Waespe W, Henn V (1977a) Neuronal activity in the vestibular nuclei of the alert monkey during vestibular and optokinetic stimulation. Exp Brain Res 27: 523–538
Waespe W, Henn V (1977b) Vestibular nuclei activity during optokinetic afternystagmus (OKAN) in the alert monkey. Exp Brain Res 30:323–330
Waespe W, Henn V (1978a) Reciprocal changes in primary and secondary optokinetic after-nystagmus (OKAN) produced by repetitive optokinetic stimulation in the monkey. Arch Psychiatr Nervenkr 225:23–30
Waespe W, Henn V (1978b) Conflicting visual-vestibular stimulation and vestibular nucleus activity in alert monkey. Exp Brain Res 33:203–211
Waespe W, Henn V (1979) The velocity response of vestibular nucleus neurons during vestibular, visual, and combined angular acceleration. Exp Brain Res 37:337–347
Waespe W, Henn V (1981) Visual-vestibular interaction in the flocculus of the alert monkey. II. Purkinje cell activity. Exp Brain Res 43:349–360
Waespe W, Henn V (1985) Cooperative functions of vestibular nuclei neurons and floccular Purkinje cells in the control of nystagmus slow phase velocity: single cell recordings and lesion studies in the monkey. In: Berthoz A, Melvill Jones G (eds) Adaptive mechanisms in gaze control – facts and theories. Elsevier, Amsterdam; pp 233–250
Waespe W, Schwarz U (1986) Characteristics of eye velocity storage during periods of suppression and reversal of eye velocity in monkeys. Exp Brain Res 65: 49–58
Waespe W, Wolfensberger M (1985) Optokinetic nystagmus (OKN) and optokinetic after-responses after bilateral vestibular neurectomy in the monkey. Exp Brain Res 60:263–269
Waespe W, Henn V, Miles TS (1977) Activity in the vestibular nuclei of the alert monkey during spontaneous eye movements and vestibular or optokinetic stimulation. In: Baker R, Berthoz A (eds) Control of gaze by brain stem neurons, Dev Neurosci, vol 1. Elsevier, Amsterdam; pp 269–278
Waespe W, Huber T, Henn V (1978) Dynamic changes of optokinetic afternystagmus (OKAN) caused by brief visual fixation periods in monkey and man. Arch Psychiatr Nervenkr 226:1–10
Waespe W, Henn V, Isoviita V (1980) Nystagmus slow-phase velocity during vestibular, optokinetic, and combined stimulation in the monkey. Arch Psychiatr Nervenkr 228:275–286

Waespe W, Büttner U, Henn V (1981) Visual-vestibular interaction in the flocculus of the alert monkey. I. Input activity. Exp Brain Res 43:337–348

Waespe W, Cohen B, Raphan T (1983) Role of the flocculurs and paraflocculus in optokinetic nystagmus and visual-vestibular interactions: effects of lesions. Exp Brain Res 50:9–33

Waespe W, Cohen B, Raphan T (1985a) Dynamic modification of the vestibulo-ocular reflex by the nodulus and uvula. Science 228:199–202

Waespe W, Rudinger D, Wolfensberger M (1985b) Purkinje cell activity in the flocculus of vestibular neurectomized and normal monkeys during optokinetic nystagmus (OKN) and smooth pursuit movements. Exp Brain Res 60:243–262

Walberg F, Bowsher D, Brodal A (1958) The termination of primary vestibular fibers in the vestibular nuclei in the cat: an experimental study with silver methods. J Comp Neurol 110:391–419

Weber JT (1985) Pretectal complex and accessory optic system of primates. Brain Behav Evol 26:117–140

Weber JT, Giolli RA (1986) The medial terminal nucleus of the monkey: evidence for a "complete" accessory optic system. Brain Res 365:164–168

Westheimer G, Blair SM (1974) Functional organization of primate oculomotor system revealed by cerebellectomy. Exp Brain Res 21:463–472

Westheimer G, McKee SP (1975) Visual acuity in the presence of retinal-image motion. J Opt Soc Am 65:847–850

Wilson VJ, Melvill Jones G (1979) Mammalian vestibular physiology. Plenum, New York

Wilson VJ, Kato M, Thomas RC, Peterson BW (1966) Excitation of lateral vestibular neurons by peripheral afferent fibers. J Neurophysiol 29:508–529

Winterson BJ, Steinman RM (1978) The effect of luminance on human smooth pursuit perifoveal and foveal targets. Vision Res 18:1165–1172

Wurtz RH, Goldberg ME, Robinson DL (1982) Brain mechanisms of visual attention. Sci American 246:100–107

Yasui S, Young LR (1975) Perceived visual motion as effective stimulus to pursuit eye movement system. Science 190:906–908

Yingcharoen K, Rinvik K (1983) Ultrastructural degeneration of a projection from the flocculus to the nucleus prepositus hypoglossi in the cat. Exp Brain Res 51:192–198

Young LR (1977) Pursuit eye movement – what is being pursued? In: Baker R, Berthoz A (eds) Control of gaze by brain stem neurons. Developments in Neurosci, vol 1. Elsevier, Amsterdam; pp 29–36

Young LR, Henn V (1975) Nystagmus produced by pitch and yaw rotation of monkeys about non-vertical axes. Fortschr Zool 23:235–246

Zee DS, Yamazaki A, Butler PH, Gueçer G (1981) Effect of ablation of flocculus and paraflocculus on eye movements in primate. J Neurophysiol 46:878–899

Rev. Physiol. Biochem. Pharmacol., Vol. 106

DC Potentials of the Cerebral Cortex

Seizure Activity and Changes in Gas Pressures

HEINZ CASPERS, ERWIN-JOSEF SPECKMANN, and ALFRED LEHMENKÜHLER

Contents

1 Introduction

The term "DC potentials" denotes bioelectrical phenomena produced by cerebral structures such as the cortex in a frequency band ranging from EEG waves to ultraslow fluctuations down to 0 Hz. The first recordings of this type of bioelectrical activity were performed using galvanometers connected to unpolarizable electrodes. By applying this technique Caton was able as early as 1875 to detect that the cerebral cortex produces a sustained potential difference superimposed with faster fluctuations. He further observed that the baseline of his recordings frequently showed slow fluctuations correlated, for instance, with sensory stimulations, motor actions, or seizure discharges (Caton 1875; cf. Brazier 1959, 1963; O'Leary and Goldring 1976; Gumnit 1974a).

Institute of Physiology, University of Münster, D-4400 Münster, FRG

The bioelectrical phenomena traced by galvanometers were easily distorted by various technical and biological artifacts produced by instabilities of electrode potentials, drifts of the recording instrument, or movements of the head. Therefore, voltage amplifiers containing high pass filters that guaranteed a stable baseline and an improved registration of high frequency potential oscillations were introduced. These recording systems eliminated information contained in slow changes of bioelectrical activity and put emphasis on higher frequency components in the range from beta to delta waves that constitute the conventional EEG.

Improved electrodes and DC amplifier systems facilitated renewed interest in recordings of slow potential changes in the cerebral cortex (Davis et al. 1939; Libet and Gerard 1941). Later on, systematic investigations of slow potential phenomena were initiated by Leao (1944, 1947), Köhler et al. (1952), and O'Leary and Goldring (cf. 1964). Since then rapid developments have occurred concerning both the recording techniques and the differentiation of biological problems. A summarizing review of the material published up to 1974 can be found in the *Handbook of Electroencephalography and Clinical Neurophysiology*, edited by A. Remond (1974).

1.1 Terminology and Definitions

The slow components of bioelectrical cortical activity recorded in various experimental situations have been given different labels at different times. Besides DC potentials, terms such as standing potentials, steady potentials, sustained potentials, and, most often, slow potential changes (SP) have been used. In one or more respects all of these labels appear unsatisfactory. The potential differences they refer to may be steady or shift slowly, while their fluctuations may also comprise high frequency components. After considering the variety of arguments for and against each particular name, it has been concluded that the term DC potentials – derived from the application of direct coupled amplifiers – is least prejudicial. Therefore, it has been recommended to give this nomenclature preference (Caspers 1974).

The term DC potential will be applied also in the present paper. Under this label will be summarized all bioelectrical events that can reliably be traced by means of a DC recording system, i.e., all events in a frequency range between 0 and ca. 100 Hz. Displacements (deviations) of the baseline will be described as DC shifts; as will be shown in a later section, the duration of such shifts can vary widely. More rapid potential fluctuations superimposed on shifts of the DC baseline will usually be characterized as "fast DC transients"; the designations deflections, fluctuations, and oscillations will be used synonymously. These labels include the waves of the conventional EEG. All indications concerning the polarity of DC shifts refer to the potential level immediately prior to the potential change. In the figures in this article negative shifts of the DC potential are represented by upward deflections, and vice versa. All tracings were recorded from the cortex against an extracerebral reference point, most often located on the nasal bone.

1.2 Scope of Material

The following presentation is primarily concerned with data obtained in animal experiments. The results of DC studies performed in man dealing, for instance, with contingent negative variations, readiness potentials, or with the late endogenous components of event-related bioelectrical phenomena have been summarized in a review by Haider et al. (1981).

The experimental results are presented in three parts. The present first part describes DC potential shifts and their correlations with neuronal, glial, ionic, and metabolic activity changes mainly on pathophysiological test conditions. In this context particular emphasis has been put on seizure activity as well as on changes in blood gas pressures (hypercapnia, hypoxia, hyperoxia), because the majority of investigations concerning DC shifts and their correlations with other functional variables have been devoted to these topics. The phenomenon of spreading depression has been excluded, since the extensive work in this special field of DC studies has already been summarized in a number of reviews and monographs (cf. Bureš et al. 1974; Nicholson and Kraig 1981; Hansen 1985).

A second part of the review of DC studies which will be published later, deals with DC phenomena associated with physiological activity changes occurring during the sleeping-waking cycle, with sensory stimulations and pharmacologically mediated damping effects. Finally, a third part will be devoted to problems of DC generator structures and electrogenesis.

2 DC Shifts Associated with Seizure Activity

2.1 Focal Seizures

Focal epileptic activity elicited by topical application of a convulsant agent or by repetitive electrical stimulations of the cerebral cortex is always associated with characteristic deviations of the DC potential (cf. Caspers 1959; O'Leary and Goldring 1964; Ayala and Walker 1965; Gumnit and Takahashi 1965; Gumnit et al. 1970; Gumnit 1979). After administration of penicillin, for example, recordings from the center of the focus at first show interictal discharges synchronized with neuronal depolarizations (Matsumoto and Ajmone Marsan 1964a, b; Sugaya et al. 1964; Matsumoto et al. 1969; Ayala et al. 1970, 1973). The development of a focus in its very beginning stages is shown in Fig. 1A, B. The tracings indicate that the predominant component of the spike consists in a negative sharp wave. With further progression of focal activity the interictal discharges frequently develop into longer periods of tonic seizures. Within a limited field around the focus such paroxysms coincide with a sustained negative deviation from the DC baseline. According to Gumnit and Takahashi (1965) the shift exhibits a maximum value in the center of the focus and decreases in amplitude, if the active recording electrode is gradually moved towards peripheral regions. Beyond a zone of complete extinction the evoked DC shifts finally reverse in sign. They correspond to some extent to the positive interictal spikes which appear at greater distances from the primary focus. Studies by Prince and Wilder (1967) as well as by Dichter and Spencer (1969a, b) on the hippocampus present evidence linking such positive changes of the field potentials to neuronal hyperpolarizations. The same conclusions can be drawn from studies by Elger et al. (1981) and Elger and Speckmann (1983), who described a zone of vertical inhibition associated with positive field potentials beneath a superficial cortical focus.

Analyses of the depth profile of the DC deviations below the center of the focus have shown that the sustained negative shifts during tonic paroxysms prevail in all cortical layers. In contrast to other bioelectrical phenomena such as sensory evoked responses, no isopotential zone seems to exist (Matsumoto and Ajmone Marsan 1964a, b; Gumnit et al. 1970); this also holds true for generalized ictal discharges (Speckmann et al. 1978). In their studies on focal convulsive activity Gumnit et al. (1970) found maximum negative DC shifts to occur at a depth corresponding to layer V. With generalized seizure attacks a comparable distribution of DC deviations is demonstrable only at the commence-

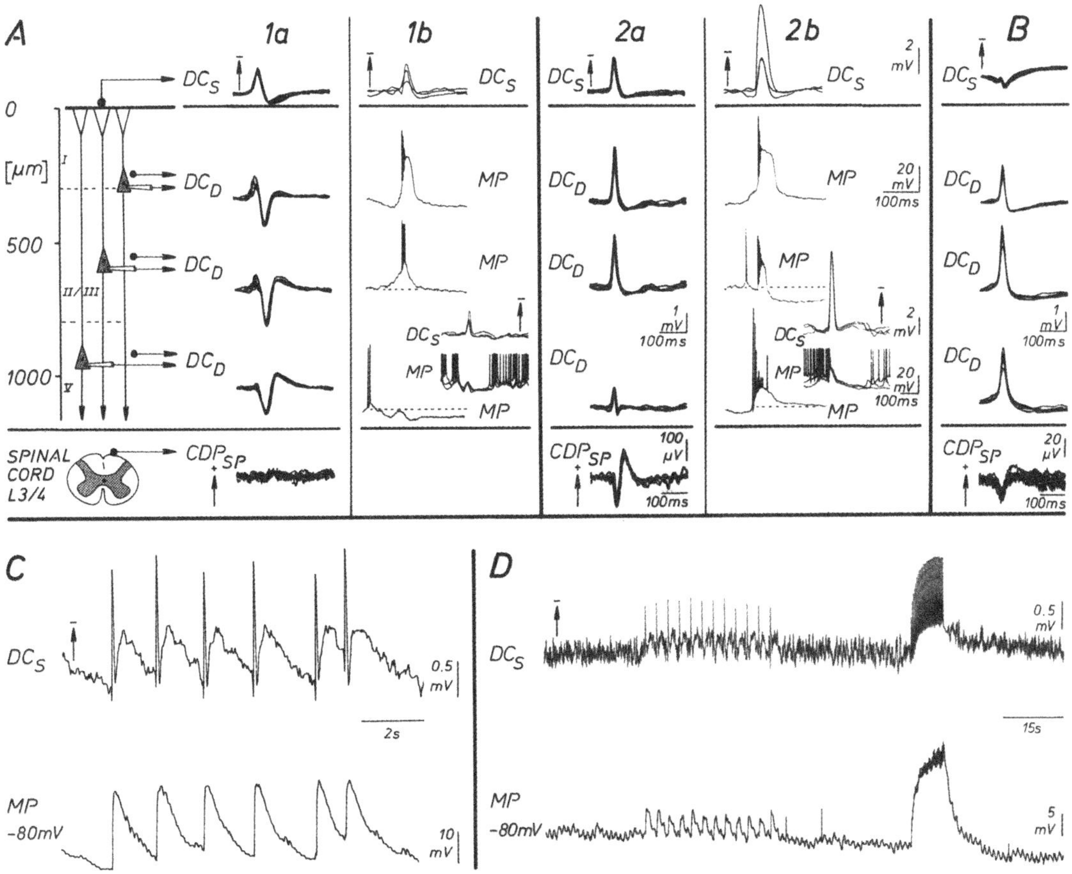

Fig. 1 A–D. Fluctuations of the cortical DC potential during focal epileptic activity and their relations to membrane potential changes of neurons (**A**) and glial cells (**C**); anesthetized and artificially ventilated rat. Epileptic discharges were elicited by local application of penicillin to the cortical surface (**A**, **C**) and to deeper cortical layers (**B**). **A** DC potential led from the cortical surface (*DCS*) and from various depths (*DCD*), neuronal membrane potential (*MP*), and the spinal cord dorsum potential (*CDPSP*) about 5 min (*1a*) and about 15 min (*2a*) after epicortical penicillin application. **B** Potential distribution following intracortical application of penicillin in layer II/III. **C** and **D** *DCS* fluctuations and *MP* changes of a cortical glial cell during focal epileptic activity (**C**) and during low-frequency and high-frequency electrical stimulation of the cortical surface (**D**). (**A** and **B** modified from Elger and Speckmann 1980, 1983)

ment of a fit. In case of prolonged ictal discharges maximum negativity is usually shifting from deeper to superficial cortical layers (see also Fig 3).

While there is a reversal of interictal spikes at greater distances from the focus, characteristic changes of the fast transients can be observed in the central area when the interictal activity is replaced by tonic paroxysms. In the beginning of such a fit the prominent negative components of the spikes sum to the sustained negative displacement of the DC baseline described above, and therefore decrease in amplitude. If the negative DC shifts surpass a critical level, the fast transients may even disappear and finally reverse in polarity (Fig. 3A). This relationship is demonstrable particularly during prolonged focal seizure attacks elicited by repetitive electrical stimulations of the cerebral cortex with low pulse frequencies (Caspers 1959). The findings suggest that the elementary mechanisms underlying the DC shifts, on the one hand, and the generation of sharp EEG waves, on the other, are closely interlinked. Supporting evidence for such an interpretation comes, moreover, from the observation that certain anticonvulsant drugs depress the development of negative DC shifts in the center of a focus while simultaneously hindering the reversal of spike polarity and the occurrence of afterdischarges (Eidelberg and Meyerson 1964; Caspers and Speckmann 1969; see also Fig. 3A). Some further details adherent to this problem will be discussed in context with generalized tonic-clonic seizures in the following section.

2.2 Generalized Seizures

Generalized seizure discharges elicited, for instance, by the systemic administration of a convulsant drug are always accompanied by distinct deviations of the DC potential all over the cortex (Fig. 2A). The extensive literature in this field has been reviewed earlier by O'Leary and Goldring (1964), Caspers and Speckmann (1969), Gumnit et al. (1970), and Gumnit (1974b, 1979). Almost all of the investigations have shown that the main component of the DC response consists in a sustained displacement of the baseline to the negative side of the preceding level. Depending on a number of factors, such as recording site, seizure type, and animal species, the predominant negative shift may be preceded, followed, and modulated by positive deflections. In this way, the configuration of DC responses associated with generalized seizure discharges can vary in details from one fit to the other. Rather constant and reproducible DC shifts appear with periodically recurring tonic-clonic convulsions elicited by repeated administration of pentylenetetrazol (PTZ) in anesthetized animals (Caspers and Speckmann 1969). Representative tracings recorded from the cerebral cortex of the rat are given in Fig. 2. The long series of almost identical DC shifts often encountered in such experimental situations offer a favorable model for special DC studies, for example, on the depth profile of the DC response, which will be dealt with in the following section.

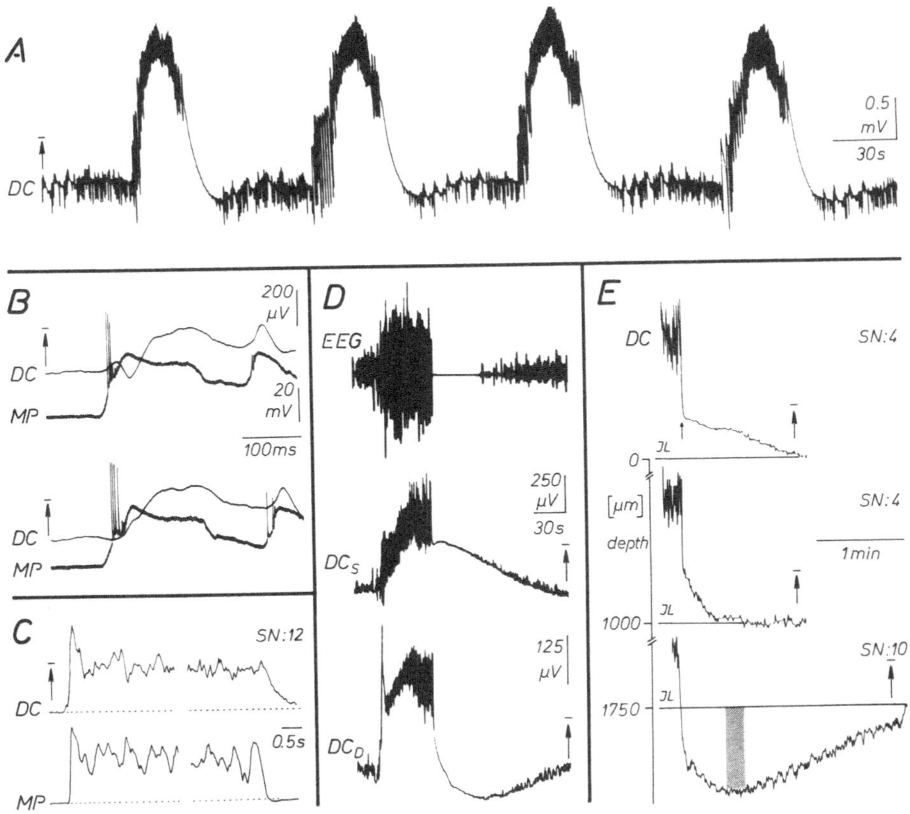

Fig. 2 A–E. Shifts and fluctuations of the cortical DC potential during generalized tonic-clonic seizure activity; anesthetized and artificially ventilated rat (**A**) and cat (**B–E**). Seizure activity was elicited by repeated injections of pentylenetetrazol. **A** DC shifts during recurrent seizures. Inkwriter recording. **B** DC potential and membrane potential of a pyramidal tract neuron (*MP*) at the commencement of two consecutive attacks. Oscilloscope recordings. **C** Average curves of *DC* shifts and of neuronal *MP* calculated from 12 consecutive attacks. **D** Conventional EEG and DC potential led from the cortical surface (DC_S) and from the depth (DC_D; 1700 µm below surface) during a tonic-clonic seizure. **E** Depth profile of cortical DC shifts at the termination (*left arrow* in the *upper curve*) of ictal episodes. The initial (preictal) DC level (*IL*) is marked by *horizontal lines*. The recording sites below cortical surface are indicated at the *left margin*. The tracings represent average curves calculated from four and ten consecutive attacks, respectively, in the same experiment. *SN*, number of sweeps. (**B**, **C**, **E** from Caspers et al. 1984; **D** from Speckmann et al. 1978)

2.3 Depth Profiles of DC Shifts

In recordings from the surface of the cerebral cortex the DC deviations associated with generalized seizure discharges frequently start with a short-lasting positive deflection (Fig. 2B). Such initial notches tend to be more pronounced in cats than in rats. As a rule, the positive wave changes rapidly to a negative DC shift which persists throughout the convulsive attack and increases in amplitude, until seizure activity stops (Fig. 2A). After cessation of the fit the negative DC deviation returns slowly to the preictal level. The decay time increases with the length of the foregoing attack and can amount to 1 min or more (Speckmann et al. 1978). The latter finding is very characteristic for surface leads.

The negative shift of the DC potential which predominates during generalized convulsive discharges in epicortical recordings can also be found to occur in deeper cortical layers. With a stepwise penetration of the cortex, this main component of the DC response does not reverse in polarity. Similar observations have been made, moreover, with other types of DC shifts. Apart from this general property, however, some particular features of the DC displacements differ in superficial and deeper cortical layers and show a clear depth profile (Speckmann et al. 1978). Thus, the positive deflections often present in surface leads and preceding the sustained negative DC shifts are usually missing in depth recordings. In the fifth cortical layer, for instance, the commencement of a convulsive attack is characterized by a unidirectional and steep negative displacement of the DC baseline. In further contrast to surface recordings this negative shift passes an early peak and then declines to a rather constant level. An example of the finding is illustrated in Fig. 2D. Another difference between DC shifts in superficial and deeper cortical layers appears at the termination of a generalized convulsive attack. As the tracings in Fig. 2D, E demonstrate, the flat decay of the evoked negative DC deviations towards the preictal level increases in slope and finally turns to a transient positive deflection, if the active recording electrode is gradually moved from the surface to deeper structures of the cortex. Such a behavior is obviously not restricted to seizure discharges elicited by PTZ, but has been found with other types of evoked DC displacements as well (Ferguson and Jasper 1971). The described difference in time course of the DC responses means that considerable potential gradients can develop between superficial and deeper cortical layers, especially towards the end and after cessation of a fit. Some consequences of this fact will be discussed in the following section.

2.4 Correlations with Fast DC Transients (EEG Waves)

The displacements of the DC baseline encountered during a generalized convulsive attack are superimposed with the spikes of the conventional EEG. In accordance with observations on focal seizure discharges, Speckmann et al. (1978) showed that time course and polarity of the fast DC transients in surface leads depend in part upon the amplitude

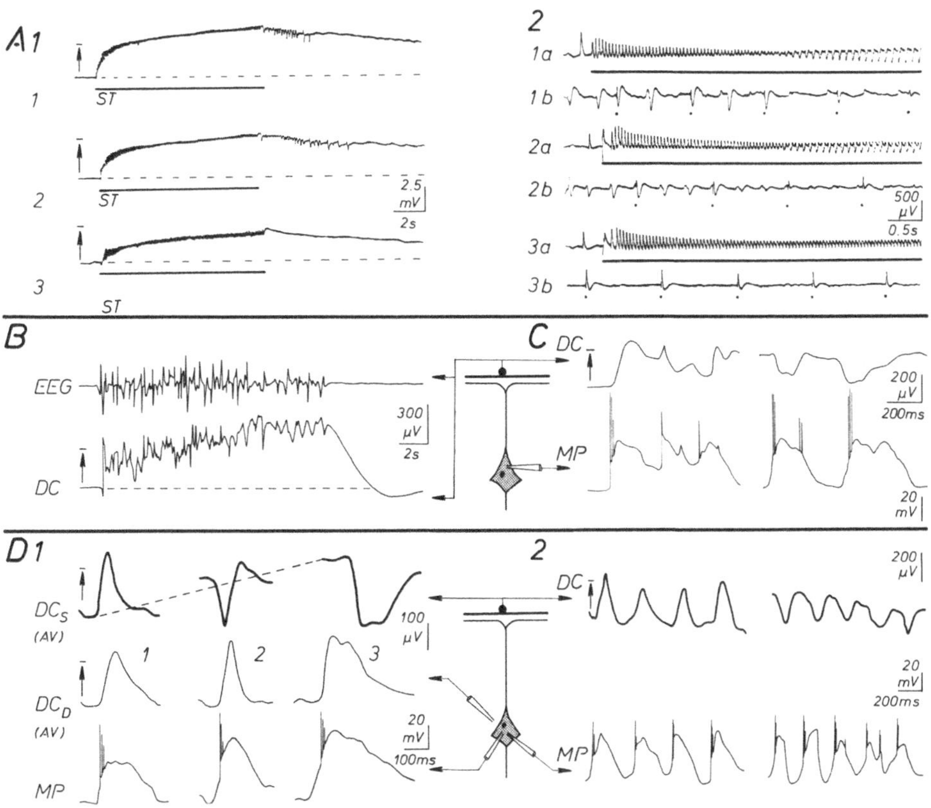

Fig. 3 A–D. Relations between polarity and shape of epicortical EEG waves and extent of negative DC shifts during high-frequency stimulation (**A**) and tonic-clonic seizure activity (**B–D**); anesthetized and artificially ventilated rat (**A**) and cat (**B–D**). Seizure activity was elicited by repeated administrations of pentylenetetrazol. **A** Epicortical potentials elicited by electrical stimulation (*ST*) and recorded by means of a DC (*1*) and AC (*2*) amplifier; different time scales. Stimulation: 20/s (*A1, A2a*) and 1/s (*A2b, dots*). In *A2b* responses to 1/s stimulation are mixed with spontaneous discharges appearing immediately after the period of high-frequency stimulation. Tracings labelled *2* and *3*: About 5 min and 45 min after administration of diphenylhydantoin. **B** Conventional EEG and epicortical DC potential during a generalized tonic-clonic seizure. **C** Simultaneous recordings of the epicortical DC potential and of the membrane potential of a pyramidal tract neuron (*MP*). **D** Single potential fluctuations at the cortical surface (DC_S), MP changes of a pyramidal tract cell (PTC) and DC fluctuations in the cortical depth (DC_D) during generalized tonic-clonic seizures. Average curves (*AV*). *D1* The negative potential (*1*), the positive-negative fluctuation (*2*), and the positive potential (*3*) in the DC_S recording coincide with monophasic negative DC_D transients and stereotyped paroxysmal depolarization shifts in the PTC. The negative DC shift occurring during the seizure is indicated by a *dashed line* in the *upper row*. *D2* The relations between DC fluctuations and MP of PTC as described for *D1* also hold true for trains of potentials. (**A** from Caspers and Speckmann 1969; **B, D** from Speckmann and Elger 1982)

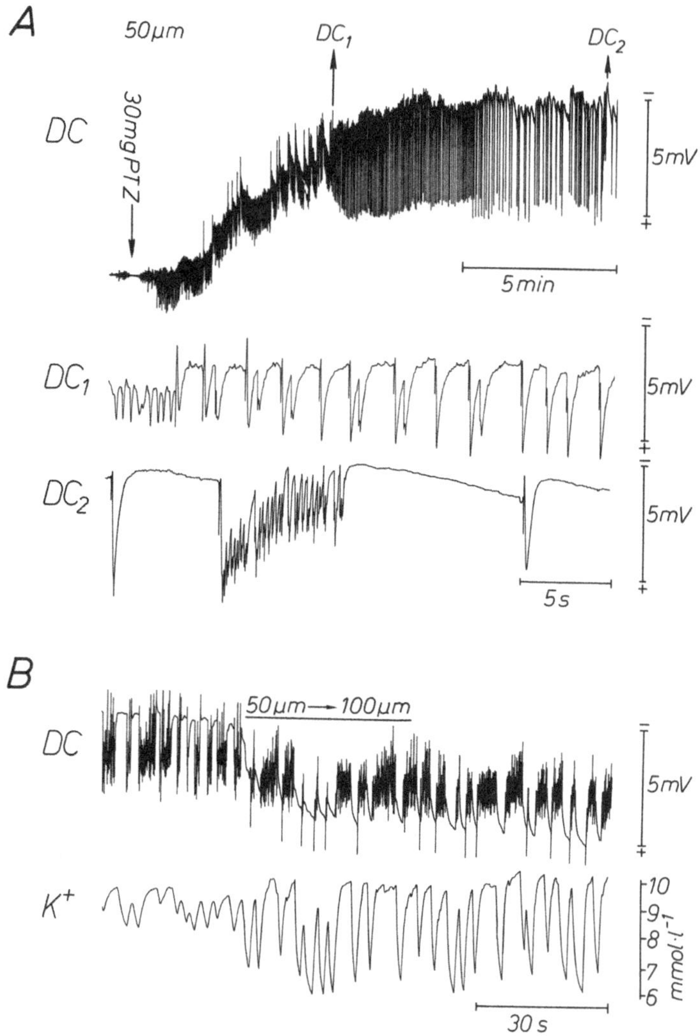

Fig. 4A, B. Displacements and superimposed fast transients of the cortical DC potential after an intraperitoneal injection of pentylenetetrazol (PTZ); anesthetized and artificially ventilated rat. **A** Recording from a depth of 50 µm below cortical surface. Parts labelled DC_1 and DC_2 are displayed with an extended time scale. **B** Simultaneous recordings of the cortical DC potential and of extracellular potassium activity (K^+). In the course of the tracing (***horizontal bar***) the double-barrelled ion selective micropipette was shifted from 50 µm to a depth of 100 µm below surface. The reversal in polarity of DC transients is associated with a reduction of the mean negative DC level. (**B** from Caspers et al. 1980)

of the sustained negative DC shift. This finding is illustrated in Fig. 3. In Fig. 3D negative, positive, and diphasic EEG spikes associated with paroxysmal depolarization shifts (PDS) in pyramidal tract cells are displayed according to the DC level on which they tend to appear. It can be seen that negative spikes occur with maximum probability at relatively low negative DC shifts, whereas positive spikes predominate at higher negative DC levels. Because the amplitude of the negative DC displacement at the surface of the cerebral cortex tends to rise in the course of a seizure, primary negative spikes may gradually decrease in voltage and finally reverse in polarity. The curves in Fig. 3D2 indicate that similar observations can be made with longer sequences of seizure patterns. Since the negative DC deviations in superficial and deeper cortical layers differ in time course, a reversal of the fast DC transients can occur, furthermore, if the active recording electrode is shifted from the surface to deeper layers of the cortex, especially towards the end of a fit. Such an experiment is illustrated in Fig. 4. It shows that the primary negative spikes turn to positive ones as the microelectrode moves from 50 μm to a depth of 100 μm. Simultaneously the DC baseline is displaced to lower negativity indicating a potential gradient of ca. 3 mV. The DC shift in Fig. 4B is reflected, moreover, by different levels of K^+ activity recorded at the same intracortical sites by means of a double-barrelled microelectrode (Caspers et al. 1980).

The relations of DC potential levels and spontaneous fast EEG transients described above are in line with earlier findings concerning the changes of the direct cortical response to a single electrical shock (DCR) during generalized seizure discharges (Caspers and Speckmann 1969). They correspond, furthermore, to the results of polarization experiments (Libet and Gerard 1941; Bishop and O'Leary 1950; Caspers 1959). These investigations showed that the negative components of various types of cortical evoked potentials decrease in amplitude and finally reverse in polarity when the recording site at the cortical surface is polarized by cathodal currents of increasing intensity. The dependency of fast DC transients upon DC levels may be derived, in part, from neuronal activity changes which will be discussed in the following section.

2.5 Correlations with Neuronal and Glial Activity

The correlations between seizure patterns in the EEG and neuronal potentials have been studied extensively mainly to assess whether and to what extent neurons contribute to the generation of field potentials in the cortex. The majority of papers devoted to this problem deal mainly with interictal convulsive discharges recorded using conventional AC amplifiers (for references, see Prince 1974). Only a limited number of investigations are concerned with sustained DC shifts associated with protracted seizure attacks (cf. Caspers and Speckmann 1969; Speckmann et al. 1972, 1978). In such experiments the DC deviations have often been led from the surface of the cerebral cortex and compared to the membrane potential (MP) of pyramidal tract (PT) cells. An example of tracings obtained under such conditions is given in Fig. 2B, C. The recordings indicate that the negative DC deflections which occur during convulsive discharges of increasing duration coincide with a series of paroxysmal depolarization shifts (PDS) in PT cells. Average curves of the two bioelectrical responses sampled from a number of consecutive fits in the same experiment show, furthermore, that the mean negative DC shift corresponds rather well to the mean depolarization of the neuron. This finding might suggest that MP changes of cells in deeper cortical layers could directly account for the generation of DC potential shifts at the cortical surface. However, such an interpretation is disproved by a more detailed evaluation of the tracings. As has already been emphasized by Speckmann et al. (1978), recordings like those displayed in Fig. 2C are not completely congruent, but differ in some essential aspects. First, the pattern of EEG spikes superimposed on the slow deviation of the DC baseline varies considerably in surface leads, whereas the correlated PDS of PT cells are rather uniform in shape. This finding which became apparent already from the tracings in Fig. 3C is illustrated in more detail in Fig. 3D1. Second, the slow decay of the negative DC shifts found in superficial cortical layers contrasts strikingly with the rapid re- and hyperpolarization of PT cells known to develop after termination of a seizure attack. If the fluctuations of the neuronal MP are compared to the DC shifts in deeper cortical layers these discrepancies are no longer apparent (Fig. 2C, D, E). The tracings in Fig. 3D show that this holds true also for single EEG spikes and neuronal PDS. In fact, the shifts of the DC potential and the MP changes of PT cells during seizure activity turn out to be congruent, provided that the DC component is recorded from immediately neighboring structures.

As a whole, these findings indicate that the slow negative deviations of the DC baseline during seizure attacks as well as the faster transients can originate, in principal, from a summation of MP changes of cortical neurons. At the same time it becomes evident that MP fluctuations of PT cells in deeper cortical layers do not essentially contribute to the generation of DC field potentials at the cortical surface.

Therefore, the problem remains that the structures and mechanisms involved in producing the DC shifts and the related spike patterns in upper cortical regions are yet to be determined. It may be anticipated that activity changes of neuronal elements play a role here, too. To test this possibility, Speckmann et al. (1978) compared DC shifts and EEG spikes occurring at the surface of the cerebral cortex during generalized seizure attacks with the discharge patterns of neurons located in upper cortical layers. The elements were identified as non-PT cells and, probably, represented relay units. The investigations showed that amplitude and time course of the negative DC shifts in surface leads are clearly reflected in the discharge frequencies of the internuncial neurons during periods of ongoing seizure activity. As a rule, a high negative DC level is accompanied by high discharge rates of the relay cells, and vice versa. Concerning the EEG spikes superimposed on the slow deviations of the DC baseline, two types of neuronal responses can be differentiated:

1. Negative spikes, which predominate at low negative DC shifts on the surface of the cerebral cortex are alays associated with a further increase in neuronal discharge rates
2. Unidirectional positive spikes, which tend to appear only at high negative DC levels, coincide with a strong reduction of neuronal discharge frequencies or, most often, with a complete blockade of the background activity. The latter statement refers to "slow" positive spikes concurring with PDS of PT cells. In addition to this special seizure pattern, sometimes "fast" positive waves can be found which arise, preferentially, from low negative DC levels in surface leads and coincide with a further increase in discharge rates of neurons in upper cortical layers. Such positive deflections resemble the primary positive components of sensory evoked potentials known to be accompanied by neuronal activation (cf. Creutzfeldt and Houchin 1974).

In summary, the available data demonstrate that both the sustained negative DC shifts and the more rapid EEG transients found in surface leads are closely related to activity changes of neurons in the superficial cortical layers. The question is whether units of the aforementioned type, supposed to be relay cells, mediate or generate the DC responses. Because systematic studies on the MP fluctuations of such elements in relation to DC shifts during generalized seizure discharges have not yet been performed, this problem remains unsettled. However, occasional intracellular recordings suggest that MP changes of neuronal perikarya in upper cortical layers cannot directly account for the observed DC shifts in surface leads. Thus, the decay of the negative DC displacement and the repolarization of neurons after the termination of a fit exhibited differences in time constants similar to those found with PT cells in layer V. In contrast to the DC responses in deeper cortical layers, which can be entirely derived from MP fluctuations of neuronal cell bodies, the

DC deviations at the surface of the cortex must have generator structures of their own. Possible candidates are dendrites and glial cells.

In recent years, a considerable experimental work has been devoted to the role played by glia in the generation of slow potential shifts (cf. Castellucci and Goldring 1970; Cohen 1974; Somjen 1975, 1979, 1984). The findings demonstrate that fast DC transients (EEG spikes) are accompanied by a depolarization of glial cells which often outlasts the EEG event (Fig. 1C). With the transition from interictal to ictal activity the single responses of the glial elements sum to a sustained depolarization that coincides with the negative shift of the DC baseline. Similar effects can be obtained with electrical stimulations (Fig. 1D). In recordings from the surface of the cerebral cortex the two bioelectrical signals are almost congruent. The correspondence disappears, particularly in the postictal period, when the active microelectrode is gradually shifted to deeper cortical layers. In contrast to the slow repolarization of glial cells, the return of the negative DC displacement becomes increasingly accelerated and is finally followed by a transient positive deflection, before returning to the initial level (cf. Fig. 2D). Comparable differences appear in depth profiles of DC shifts and changes in potassium activity, as described in the following section.

2.6 Correlations with Ionic Activity Changes

The shifts of the cortical DC potential during seizure activity described in the preceding sections are always associated with changes of ion activities in the extracellular space. In this context especially K^+, Ca^{2+}, Na^+, Cl^-, and H^+ have been studied (cf. Prince et al. 1978; Lux and Heinemann 1983; Somjen 1979; Caspers et al. 1980, 1984; Dietzel et al. 1982b; Lehmenkühler et al. 1982a, b; Lux et al. 1986). However, but a few of these investigations are primarily concerned with the correlations between DC potentials and ion activities, whereas the majority of the experiments focus on the ionic response while the bioelectrical signal, if recorded, served as a mere indicator of the seizure discharge. Therefore, of the extensive literature published on changes of ion activities during seizure discharges, only a limited amount can be considered in this context. As a rule, measured ionic activities are quantified and expressed as molar concentrations of free ions in the extracellular space.

2.6.1 Changes in $[K^+]_O$

Most of the studies concerning DC potentials and ion activities in seizure conditions deal with changes in $[K^+]_O$. They demonstrate that interictal spikes elicited, for instance, by topical application of penicillin always coincide with an increase in $[K^+]_O$. In a medium range the rise of K^+ activity above the resting level of 3 mmol/l is positively correlated with the amplitude and duration of the bioelectrical signal. Similar observations have been made with primarily generalized interictal discharges

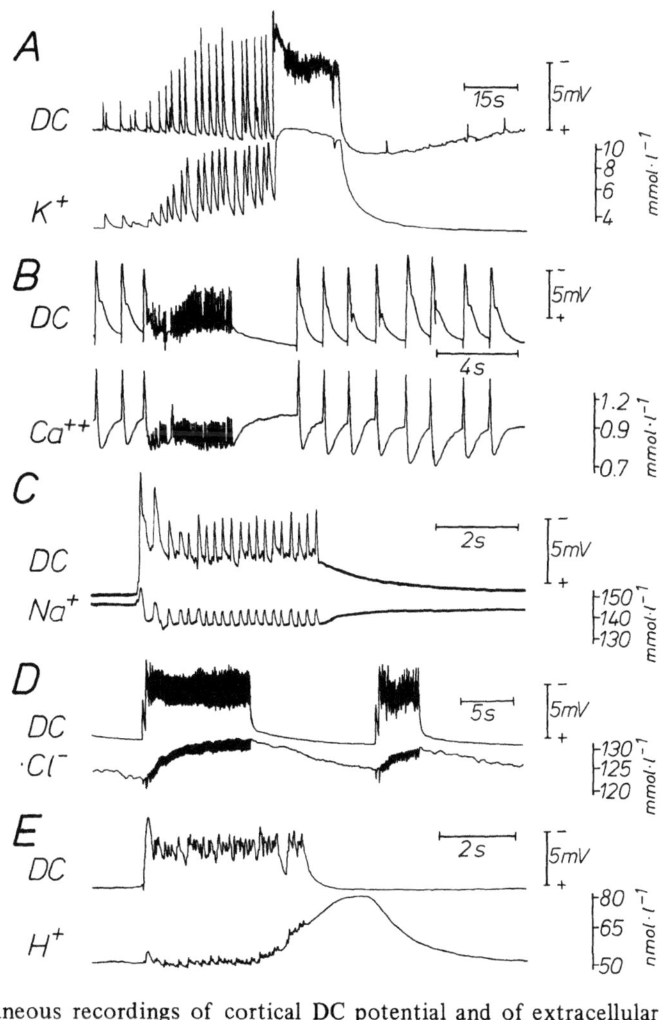

Fig. 5 A–E. Simultaneous recordings of cortical DC potential and of extracellular activity of potassium (K^+), calcium (Ca^{2+}), sodium (Na^+), chloride (Cl^-), and hydrogen (H^+) ions during seizure activity in upper cortical layers of the rat's cerebral cortex. Epileptic activity was elicited by repeated intraperitoneal injections of pentylenetetrazol in the anesthetized animal

induced by systemic administration of various convulsant drugs. At the transition from interictal to ictal activity the increase of spikes frequency, as has already been pointed out in a previous section, is usually accompanied by a shift of the DC baseline towards the negative side of the preceding level with the amplitudes of the individual spikes being simultaneously reduced. The tracings in Fig. 5 indicate that $[K^+]_O$ behaves in a very similar way. With rising spike frequency in the bioelectric recordings, the phasic increases in $[K^+]_O$ sum to a sustained elevation of the baseline, the magnitude of which depends on the actual discharge rate. Studies by Hotson et al. (1973), Heinemann and Lux (1977), and others have provided evidence that the rise in $[K^+]_O$ encountered in such experimental situations usually shows a ceiling effect at concentrations of about 12 mmol/l. Comparable response characteristics can principally be detected also in recordings of the DC potential and appear most pronounced if the shifts from the DC baseline originate in a narrow cortical field. On the whole, most of the available data indicate a close correlation of DC shifts and changes in $[K^+]_O$ during seizure activity (cf. Hotson et al. 1973; Prince et al. 1973; Lux 1974; Moody et al. 1974; Heinemann and Lux 1977; Somjen 1979; Heinemann et al. 1979; Caspers et al. 1980).

The conclusions cited above refer to recordings from the surface and/or from superficially located structures of the cerebral cortex. Since the DC shifts associated with seizure activity exhibit a characteristic depth profile, the question arises as to the correspondence between the bioelectrical and ionic signals in different cortical layers. In superficial structures, on the one hand, the negative DC shifts occurring during seizure activity are usually marked by a gradual increase followed by a slow decay in the postictal phase. In deeper layers, on the other hand, they peak early, decline to a considerably lower level, and, finally, turn to a transient positive deflection after seizure discharges have ceased. As a result, with short-lasting fits the maximum negative displacement of the DC potential appears in the depth of the cortex, whereas it moves to the surface when seizure discharges continue beyond 30–60 s. The changes in $[K^+]_O$ occurring during ictal activity show a corresponding, time-dependent depth profile. With interictal spikes as well as with short-lasting ictal episodes the maximum rise of $[K^+]_O$ has been found to occur in deeper cortical structures (Heinemann et al. 1979; Caspers et al. 1980). Thus, in the rat cortex the peak values of both the ionic and the bioelectrical signals associated with single interictal spikes regularly appear at a depth of 500–700 μm below surface (cf. Fig. 6). With longer-lasting seizure activity, the maximum increase in $[K^+]_O$ shifts from the deeper to superficial cortical layers and, thus

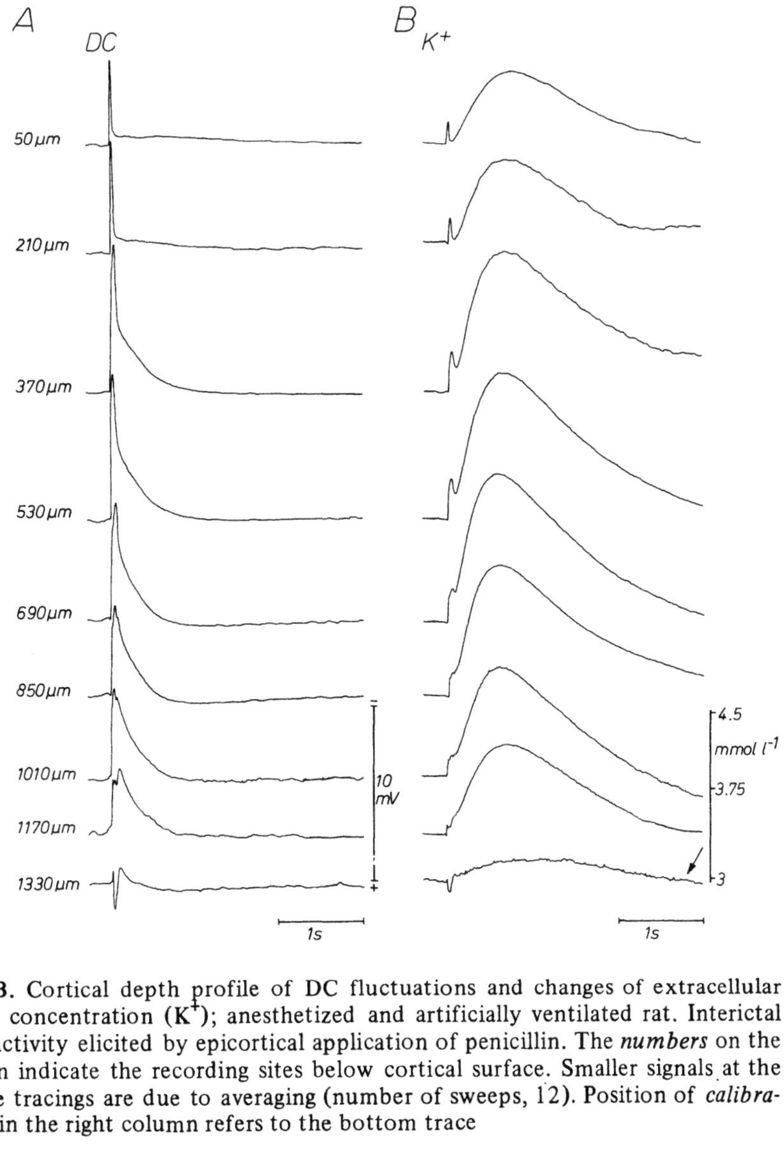

Fig. 6A, B. Cortical depth profile of DC fluctuations and changes of extracellular potassium concentration (K^+); anesthetized and artificially ventilated rat. Interictal epileptic activity elicited by epicortical application of penicillin. The *numbers* on the left margin indicate the recording sites below cortical surface. Smaller signals at the end of the tracings are due to averaging (number of sweeps, 12). Position of *calibration scale* in the right column refers to the bottom trace

again, mirrors the behavior of the DC potential. In the latter case a stepwise penetration of the cortex by the recording microelectrode results in a positive displacement of the DC baseline with the level of K^+ activity simultaneously being reduced. Such a finding is shown in Fig. 4B. The tracings furthermore demonstrate that the fast DC transients (EEG spikes) superimposed on the slow deviations of the baseline reverse in polarity when the recording electrode is shifted from superficial to deeper cortical layers. This finding is in agreement with earlier observations which show a progressive reduction and final reversal of fast negative EEG spikes occurring with increasing slow negative shifts of the DC baseline at the same recording site (Caspers 1959; Caspers and Speckmann 1969; Speckmann et al. 1978). The congruence of negative DC shifts and changes in $[K^+]_o$ found throughout a convulsive attack is lacking in the postictal phase. As a rule, the return of $[K^+]_o$ to the preictal level outlasts the duration of the DC deviation. In the depth profile of the cortex this difference increases from the superficial to deeper layers (Caspers et al. 1984). Concerning correlations with DC displacements, the changes in K^+ activity are opposite to those of neuronal membrane potentials, which correspond to the final DC deflections more precisely in depth recordings.

The relations of DC potentials and $[K^+]_o$ described so far refer, preferentially, to baseline shifts in the negative direction. The question arises as to the behavior of $[K^+]_o$ in the course of a primary positive DC displacement which can be differentiated from the decline of a preceding negative one. Fast positive DC transients have been found to coincide with interictal discharges of a cortical focus, predominantly in a surrounding zone of inhibition both in the horizontal dimension (Gumnit and Takahashi 1965; Prince and Wilder 1967; Gumnit et al. 1970) and in deeper cortical layers (Elger et al. 1981; Elger and Speckmann 1983). With a transition of focal activity from interictal to ictal discharges, the resulting shifts of the DC baseline recorded from the same peripheral site consist of an initial positive deflection followed by a slowly increasing negative deviation. In both instances the primary positive DC responses are associated with a moderate rise of $[K^+]_o$ which differs from that during negative waves only in a quantitative aspect. This finding is at variance with the decrease of K^+ activity observed by Lux (1974) and Heinemann et al. (1979) during primary positive waves evoked in the cortex by thalamic stimulation.

2.6.2 *Changes in* $[Ca^{2+}]_o$

In the literature the reference level given for $[Ca^{2+}]_o$ in the cerebral cortex varies, probably depending on the HCO^-_3 contents of the calibration solution. In our experiments normal values ranged from 0.9 to 1.1 mmol/1. A slight fall of cortical $[Ca^{2+}]_o$ to below this level may occur already in an early stage of a developing fit before definite seizure discharges appear (Heinemann et al. 1977; Somjen 1980; Pumain et al. 1983). This finding corresponds to the slow negative DC deviations elicited, for example, by systemic administration of pentylenetetrazol (PTZ) and frequently preceding the onset of a convulsive attack (Caspers and Speckmann 1969). With the occurrence of interictal discharges,

$[Ca^{2+}]_O$ shows additional variations associated with each single spike. The main component of the response consists in a further decay (Heinemann et al. 1977, 1978). Provided that electrodes of a sufficiently short response time are applied, a fast rise of $[Ca^{2+}]_O$ may be observed preceding the slower decline (Labeyrie and Koechlin 1979; Lehmenkühler et al. 1979; Pumain et al. 1983). Correspondingly, the changes of $[Ca^{2+}]_O$ associated with ictal activity often show a fast initial increase followed by a fall of up to 0.3 mmol/l (Fig. 5B). With prolonged seizure discharges the lowering of $[Ca^{2+}]_O$ is usually diminished and may even change to a secondary rise that overshoots the reference level during the postictal period (Heinemann et al. 1977, 1978, 1981, 1986; Somjen 1980; Somjen et al. 1986; Caspers et al. 1980). In depth profiles determined in the rat cortex, the changes of $[Ca^{2+}]_O$ appear similar in all layers, with some quantitative differences concerning amplitude and time course of the various response components. Thus, the initial rise of $[Ca^{2+}]_O$ reaches a maximum value at a depth of 300 μm. Moreover, the reincrease of $[Ca^{2+}]_O$ during longer lasting seizure activity and the subsequent overshoot in the postictal period become more pronounced when the recording microelectrode is gradually shifted from superficial to deeper layers. In this respect the depth profile of the ionic signal resembles that of the DC potential. A further correspondence between DC displacements and changes in $[Ca^{2+}]_O$ is encountered with seizure attacks exceeding a duration of 1–2 min. In this condition the maximum amplitude of the negative DC response shifts from deeper to superficial cortical layers (Fig. 2D). At the same time the lowest level of $[Ca^{2+}]_O$ moves towards the surface, while $[K^+]_O$ in the same direction tends to rise (Fig. 4B).

The changes of $[Ca^{2+}]_O$ described so far refer to seizure discharges associated with negative DC deviations. Few observations have been made with the bioelectrical signal shifting in the opposite direction. They suggest that in the course of a fit the reactions of $[Ca^{2+}]_O$ during positive DC deflections are not essentially different from those during negative ones. In this respect $[Ca^{2+}]_O$ apparently behaves in the same way as $[K^+]_O$.

2.6.3 Changes in $[Na^+]_O$ and $[Cl^-]_O$

The relations between DC shifts and changes in $[Na^+]_O$ and $[Cl^-]_O$ have been studied during seizure activity evoked by systemic administration of PTZ (Lehmenkühler et al. 1982a) and during electrically induced afterdischarges (Dietzel et al. 1982a, b). The results showed that interictal DC transients (EEG spikes) elicited by PTZ usually coincide with changes in $[Na^+]_O$ composed of an initial fast rise and a subsequent slow decline.

Depth profiles of the response pattern determined in the rat cortex demonstrate that the two components appear in all layers reaching maximum amplitudes at 600–800 μm below the surface. At all recording sites the initial rise of $[Na^+]_o$ proves independent of voltage and polarity of the concurrent bioelectrical signal. With the transition from interictal to ictal seizure discharges the slow components of the interictal responses sum to a sustained shift of $[Na^+]_o$ by 7–10 mmol/l below the normal level of 145–150 mmol/l. As a rule, the baseline remains superimposed with phasic increases synchronized with each single spike (Fig. 5C). As described by Dietzel et al. (1982), $[Na^+]_o$ in the deeper cortical layers tends to return to baseline in the course of longer lasting fits. This finding to some extent corresponds to the behavior of the local DC potential (cf. Fig. 2D).

The changes of $[Cl^-]_o$ during interictal fast DC transients (EEG spikes) elicited by systemic administration of PTZ consist in a small decline followed by a slow and flat rise. In a depth profile determined in the rat cortex both components exhibit maximum amplitudes at 700–800 μm below the surface. In the course of ictal activity the flat increases of $[Cl^-]_o$ associated with each single spike sum to a sustained rise by ca. 5 mmol/l above the initial level of 125–130 mmol/l. An example of this finding is presented in Fig. 5D. Occasionally diphasic responses of $[Cl^-]_o$, most often composed of a primary rise and a secondary fall, have been observed. The changes of $[Na^+]_o$ and $[Cl^-]_o$ described above refer to seizure activity associated with negative shifts of the DC potential at the corresponding recording site. Systematic studies on the relations between the ionic and bioelectrical signals during positive DC shifts are at present obviously lacking.

2.6.4 Changes in $[H^+]_o$

The correlation of DC shifts and changes in $[H^+]_o$ during ictal seizure discharges have been studied by Caspers and Speckmann (1972) using pH electrodes with tip diameters of 2 mm applied to the surface of the cerebral cortex. Their investigations showed that the negative displacement of the cortical DC potential during a seizure attack is usually associated with an alkaline or biphasic alkaline-acidic shift at the cortical surface, if the measurements are performed at the very beginning in a series of subsequent fits. In such cases the initial decrease of $[H^+]_o$ always coincides with a considerable rise of cortical blood flow. With repetitive ictal discharges the phasic increases in blood flow rapidly flatten with the amplitude of the alkaline shift simultaneously being reduced. Finally, each seizure attack is accompanied by a unidirectional acidic response.

These earlier measurements have been supplement by studies using ion-selective microelectrodes to determine seizure-bound $[H^+]_O$ shifts in different cortical layers. The findings at first confirm that the main component of the $[H^+]_O$ reaction during ictal activity consists in an acidic shift regularly preceded by an alkaline wave (Urbanics et al. 1978; Lehmenkühler et al. 1982b; Kraig et al. 1983). However, the initial decrease of $[H^+]_O$ exhibits a clear depth profile. It is most prominent in recordings from superficial cortical layers and disappears in those from deeper structures. An original recording of the latter reaction is presented in Fig. 5E. As a whole, no significant correlation between DC potential shifts and changes in $[H^+]_O$ during sustained seizure activity is discernible. This statement refers to recordings from the surface as well as from deeper layers of the cortex.

2.6.5 Changes of the Extracellular Space

To estimate possible contributions of transmembranous water flux to changes of extracellular ion activities, the fraction of the extracellular space (ES) with respect to the total brain volume has been measured. Using the tetramethylammonium (TMA^+) iontophoresis technique (cf. Neher and Lux 1973; Nicholson and Phillips 1981), Lehmenkühler et al. (1985) determined that the ES occupies about 20% of the total volume in all layers of the rat cortex. Similar results have been obtained by Nicholson and Phillips (1981) in the cerebellum. During ictal seizure discharges induced by electrical stimulation in the cortex of cats Dietzel et al. (1980, 1982a, b) and Lux et al. (cf. 1986) found on average a 30% shrinkage of the ES in comparison with the initial value, with the maximum located in a deeper cortical layer. Transient decreases of ES ranging from 0.5% at a depth of 600 μm to 3% at 1500 μm have also been observed in connection with individual interictal spikes (Lehmenkühler et al. 1986).

3 DC Shifts Associated with Changes of Gas Pressures in Blood and Tissue

3.1 Changes of pCO_2 and/or pH

Changes in arterial pCO_2 have been found to exert a considerable influence on cortical DC potentials (Goldensohn et al. 1951; Tschirgi and Taylor 1958; Libet and Gerard 1962; Caspers 1963, 1965; Caspers et al. 1963; Caspers and Speckmann 1969, 1971; O'Leary and Goldring 1964;

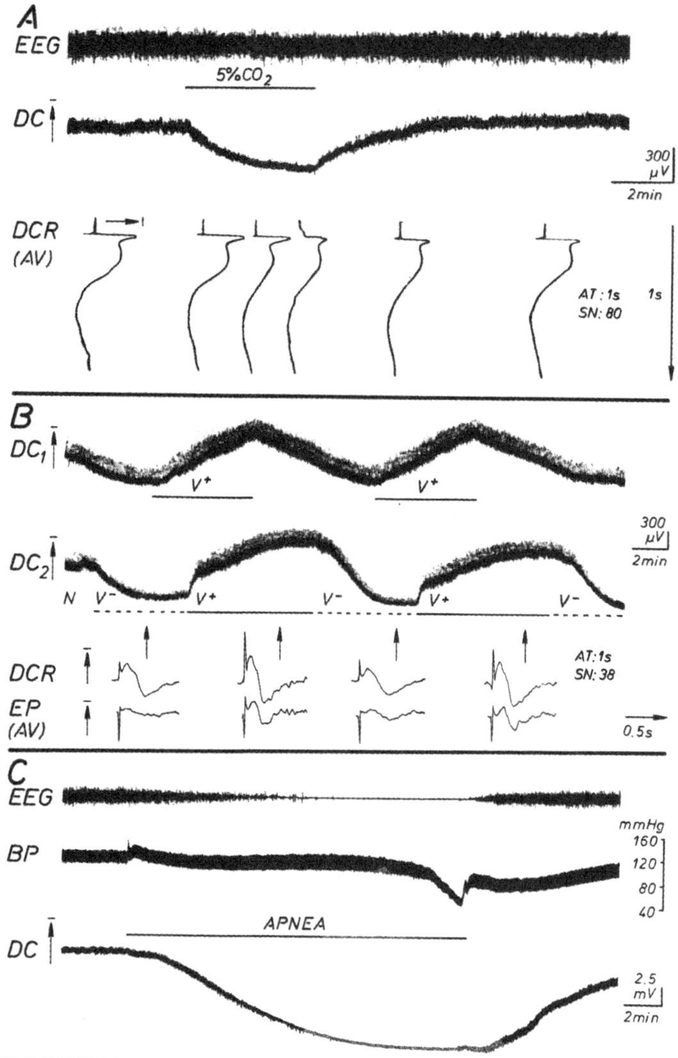

Fig. 7 A–C. Shifts of the cortical DC potential elicited by changes of the CO_2 pressure in tissue; anesthetized and artificially ventilated rat. **A** Addition of 5% CO_2 to the inspired gas mixture (21% O_2, 74% N_2; *horizontal bar*). The conventional EEG and the averaged direct cortical response to a single electrical shock (*DCR, AV*) were recorded simultaneously. Analysis time (*AT*); 1 s; number of sweeps (*SN*), 80. **B** Changes of the ventilation rate (*V*). DC_1 and DC_2: Two different experiments. V^+ (*solid lines*), hyperventilation; V^- (*dashed lines*), hypoventilation; *N*, normal ventilation. Averaged DCR to single electrical shocks and transcallosal evoked potentials (*EP, AV*) were recorded simultaneously and related to DC_2 by *arrows*. *AT*, 1 s; *SN*, 38. **C** Respiratory arrest following a longer period of breathing pure oxygen (apnea, *horizontal bar*). The conventional EEG and the arterial blood pressure (*BP*) were recorded simultaneously. (**A** and **B** from Speckmann and Caspers 1969; **C** from Caspers et al. 1979)

Besson and Aléonard 1965; Besson et al. 1970; Speckmann and Caspers 1967, 1969; Wurtz and O'Flaherty 1967; Hirsch et al. 1968a, b; Woody et al. 1970; Loeschcke 1956, 1971). Apart from this general statement there is, however, little consensus on polarity and time course of the evoked DC shifts. Thus, experimental observations have been published suggesting opposite DC responses to a rise of the pCO_2 in different animal species (O'Leary and Goldring 1964; Woody et al. 1970; Loeschcke 1971). Our own experiences, gathered in investigations using rats, showed that in these animals the proper reaction of the cortical DC potential to an increase in pCO_2 consists in a unidirectional baseline shift to the positive side of the preceding level, and vice versa. The findings in rats will be described first, contrary observations will be considered in a following section.

In the anesthetized rat, artificially ventilated at a constant rate, a rise of the inspiratory pCO_2 always evokes a positive DC shift. An example of this finding is presented in Fig. 7A. Principally the same type of DC response occurs in a variety of other experimental situations giving rise to an increase in pCO_2. As shown in Fig. 7B, each decrease of the ventilation rate is accompanied by a positive DC deflection, and vice versa. A further experimental procedure to test the reactions of the DC potential to an increase in pCO_2 is illustrated in Fig. 7C. It consists in respiratory arrest following a longer period of breathing pure oxygen with the lungs kept connected to the O_2 container. In this condition the pCO_2 rises to excessive values with the tissue pO_2 ranging above normal levels (cf. also Fig. 8B; Caspers and Speckmann 1969, 1971, 1974; Speckmann et al. 1970). In contrast to asphyxiation, this type of respiratory arrest, which will be referred to as "apnea", is consistently associated with a positive shift of the DC potential. With all of the positive DC shifts elicited by the various procedures described above no characteristic depth profile of the response is discernible. As a rule, the positive shifts are demonstrable throughout the cortex, with no reversal in polarity taking place.

In either case the evoked shifts of the cortical DC potential are clearly related to changes in arterial pCO_2 and/or pH. At least in a medium range of pCO_2 deviations from normal levels, the voltage of the DC shifts has been found to depend approximately linearly on the changes in pH and in the log pCO_2 (Caspers and Speckmann 1974). This finding allows the prediction that the amplitude of each DC shift induced, for instance, by fluctuations of the ventilation rate, is determined by the preexisting CO_2 pressure level. Consistent with the shape of the CO_2 dissociation curve, a definite reduction of the respiratory volume should elicit considerable positive DC deflections, if it starts

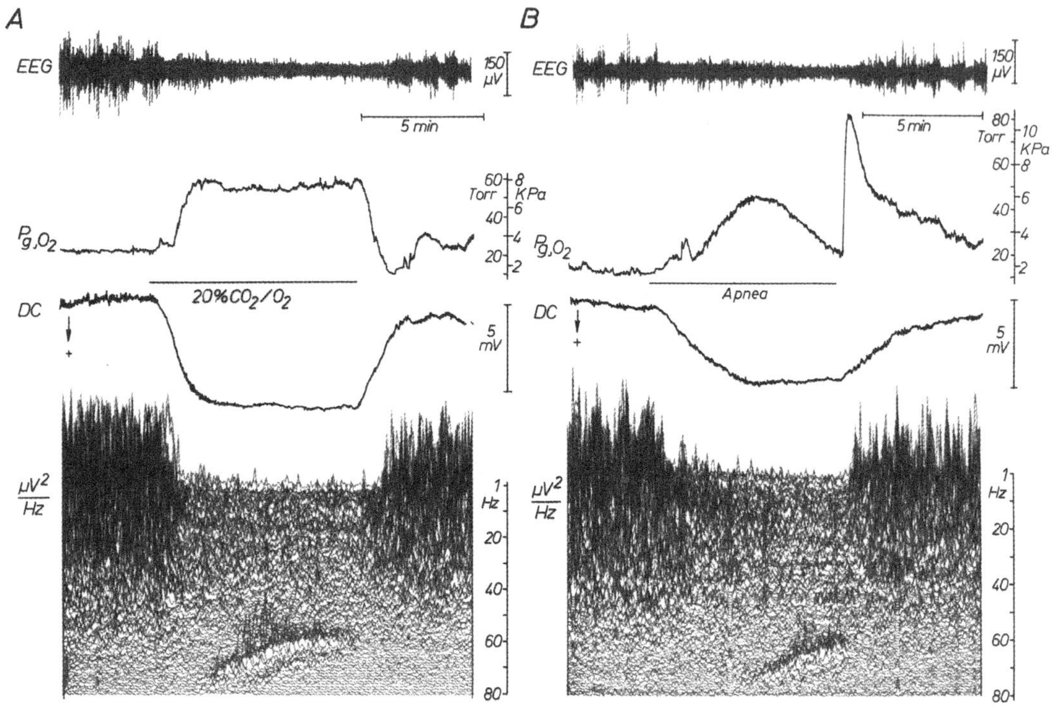

Fig. 8A, B. Relations between cortical DC shifts and changes of the conventional EEG, of the oxygen pressure in tissue (P_g, O_2) and of the power spectrum of the EEG (**A**) during elevation of the inspiratory CO_2 content (20% CO_2/O_2, *horizontal bar*) and (**B**) during respiratory arrest after ventilation with pure oxygen for about 30 min (apnea, *horizontal bar*); anesthetized and artificially ventilated rat. High frequency EEG waves appear transiently during the increase of the CO_2 pressure in tissue

from higher values of the pCO_2. With preceding hyperventilation, however, the same diminution of the ventilation rate should be much less effective. These predictions have actually been confirmed (Caspers and Speckmann 1974). Such relations could be responsible, in part, for the observed variability in amplitudes of DC responses to changes in pCO_2. The question may arise whether the evoked DC displacements can actually be attributed exclusively to changes in pCO_2 and/or in pH. As a rule, the described changes in pCO_2 in either direction are accompanied also by fluctuations of the pO_2, the amount as well as the polarity of which depend on a number of mechanisms such as oxygen uptake in the lungs, reactions of cortical blood flow, and local oxygen consumption, and are therefore hard to predict. It has been shown, however, that variations of the pO_2 in a medium range exerted no measurable influence on the DC potential level. Such a conclusion can be drawn, for instance, from the observation that a positive DC displacement during experimental hypoventilation occurs independent of whether tissue pO_2 in the cerebral cortex increases or declines. Interactions between CO_2 and O_2 effects do not become evident before the fluctuations of the pO_2 exceed a lower or upper limit (Caspers and Speckmann 1974).

The actions of CO_2 on cortical DC potentials described above are generally agreed upon as far as experiments on rats are concerned. The same statement holds true for studies performed on dogs (Goldensohn et al. 1971; Hirsch et al. 1968a). Similar DC deviations have been observed by Wurtz and O'Flaherty (1967) also in cats, but this is the animal for which findings are controversial. Thus, Libet and Gerard (1962) described positive DC deviations elicited by hyperventilation. In more recent studies Woody et al. (1970), Besson et al. (1970), and Carpenter et al. (1974) reported that respiratory acidosis and hypercapnia were usually accompanied by negative deflections of the cortical DC potentials both in cats and monkeys. The question arises whether such opposing DC responses depend on primary differences of elementary CO_2 effects in different animal species or on secondary mechanisms mediated, for instance, by changes in local pO_2 and/or cerebral circulation.

A number of comparative studies performed by Caspers and Speckmann (1974) and Caspers et al. (1979) both on rats and cats contribute to a further clarification of the problem. Their results showed that a rise of the pCO_2 can elicit negative DC shifts also in rats if carbon dioxide is administered dissolved in oxygen. In such a case the evoked increase in tissue pO_2 may exceed a critical level at which O_2-dependent negative DC shifts appear (cf. Fig. 18). Furthermore, it was found that time course and polarity of CO_2-induced DC reactions are related to fluctuations of the brain volume and, hence, of the intracranial pressure. Such a dependency was demonstrable particularly in studies on cats. Also in this animal the evoked DC shifts turned out to be positive in polarity, if a remarkable increase in brain volume during administration of CO_2 was missing. With increased swelling of the brain, however, the initially positive DC response was replaced by a negative one. These findings demonstrate that both hyperoxia and hypoxia represent possible factors inducing negative DC shifts associated with hypercapnia. They suggest that the proper DC response to an increase in pCO_2 consists in a positive shift – independent of the animal species concerned. This interpretation corresponds, moreover, to the observation that a rise of the pCO_2 is quite generally effective to prevent, interrupt, or delay negative DC shifts that occur during seizure activity (Caspers and Speckmann 1969, 1972), cortical spreading depression (Caspers and Speckmann 1974; Lehmenkühler et al. 1976), and oxygen deficiency (Caspers and Speckmann 1971, 1974).

The statement that positive DC shifts during hypercapnia represent the proper DC response to an increase in pCO_2 refers, preferentially, to studies on cortical structures. It does not exclude that true negative DC deviations may arise somewhere in the brain, where excitatory actions of CO_2 or generator structures other than neurons predominate (Loeschcke 1971; Tschirgi and Taylor 1958).

3.1.1 Relations with EEG Waves

The positive shifts of the cortical DC potential elicited by a rise in pCO_2 are always associated with changes of both spontaneous and evoked EEG waves (cf. Withrow 1972; Caspers and Speckmann 1974; Caspers et al. 1979). The most prominent feature of the effects of CO_2 on spontaneous cortical activity is the well-known depression of EEG potentials in the conventional frequency band. As a rule, slow potential fluctuations appear to be more susceptible than faster waves. Such effects are particularly demonstrable by means of the power spectra displayed in Fig. 8. This evaluation shows that high frequency waves of 50–80 Hz appear transiently, if the rise of the pCO_2 passes a range of 70–130 mmHg. The depressant actions of CO_2 reflected in the changes of spontaneous EEG waves have also been demonstrated with a variety of cortical potentials evoked by direct, afferent and sensory stimulations (Fig. 7A, B). In this case, too, the slow late response components prove more susceptible than the sharper early deflections. Finally, however, even steep and highly synchronized potential fluctuations, such as seizure potentials, become regularly extinguished when the arterial pCO_2 is raised to a critical level.

3.1.2 Correlations with Neuronal and Glial Activity

In all cortical neurons studied so far a rise of the pCO_2 led to a number of common responses independent of the actual localization of the unit in different cortical structures. As a rule, CO_2 caused an increase in resting membrane potential (cf. Carpenter et al. 1974; Caspers et al. 1986). A representive finding obtained with a pyramidal tract cell in the cat cortex is illustrated in Fig. 9A. The same hyperpolarizing response has been shown to occur with the majority of spinal neurons (Washizu 1960; Caspers and Speckmann 1972; Speckmann and Caspers 1974). These studies furthermore demonstrated that antidromically evoked action potentials are abruptly suppressed when the pCO_2 is continuously

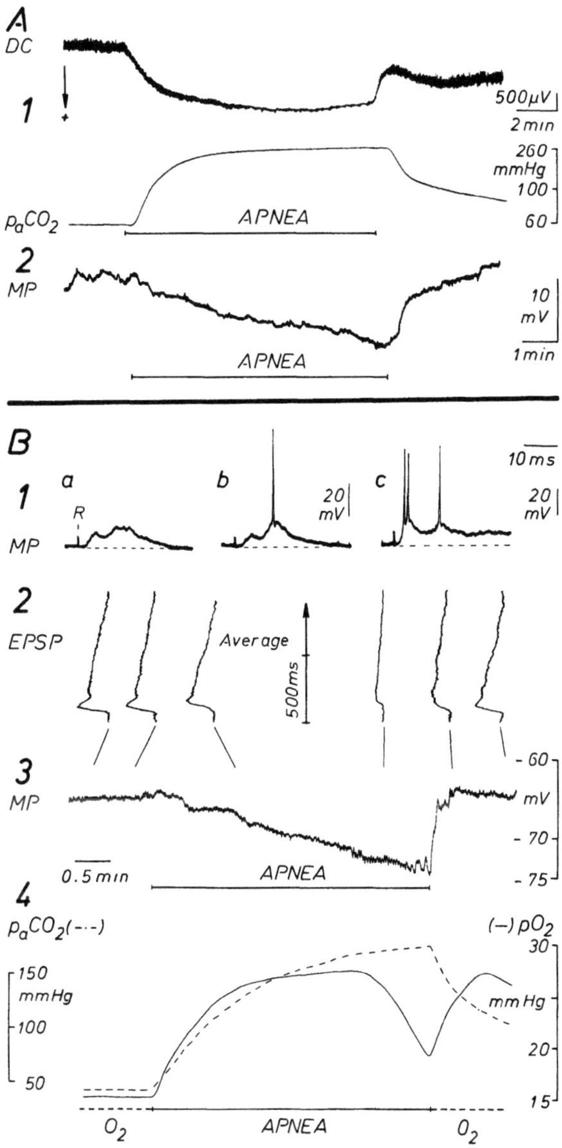

Fig. 9A, B. Relations between cortical DC shifts and changes in neuronal potentials evoked by a rise in the (arterial) p_aCO_2 during ventilatory arrest in the anesthetized rat after breathing 100% O_2 for about 15 min (apnea). With this technique the general elevation of the pCO_2 in blood and tissue is accompanied by an increase in tissue pO_2 (*B4*). **A** Positive shift of the cortical DC potential (*1*) and increase in membrane potential (*MP*) of a single neuron (*2*) elicited by a rise in the pCO_2 (period of apnea). Inkwriter tracings. **B** Increase in MP of a spinal motoneuron (*3*) during a period of apnea. The changes of arterial pCO_2 and of tissue pO_2 at the recording site are displayed in *4*. In addition to the MP the excitatory postsynaptic potentials (*EPSP*) evoked by stimulation of the motor cortex with single electrical pulses were measured and averaged throughout the experiment (*2*). The relations to the MP shifts are indicated by ***black lines***. The curves in **1** represent original oscilloscope recordings of the postsynaptic responses to stimuli (*R*) of increasing intensity (*a, b, c*). (From Caspers, Speckmann, and Lehmkühler 1984)

raised. At the same time repetitive neuronal discharges elicited by depolarizing intracellular current injections become progressively reduced, thus turning from a tonic to a phasic type of response. This phenomenon fits with the damping effects of CO_2 known from studies on peripheral nerves (cf. Monnier 1955). Besides these actions pointing to a direct influence of CO_2 on neuronal membranes, interferences with synaptic transmission processes are evident. When the pCO_2 is raised, excitatory and, most probably, also inhibitory postsynaptic potentials decrease in amplitude with the membrane resistance often being enhanced. As a result, impulse transmission is rapidly cut down. Neurons showing this type of CO_2 response have been labeled as I neurons (Speckmann et al. 1970).

In contrast to the I type cells, other units labeled E neurons have been found to be activated by CO_2 (Speckmann et al. 1970). Up to now this type of cell has been shown to occur in limited numbers only in extracortical structures. Thus, in studies on the spinal cord not more than approximately 5% of spontaneously active neurons turned out to be of the E type. Furthermore, also the CO_2-sensitive peripheral and central chemoreceptors may belong to this group.

The response of the E neurons to an increase in pCO_2 resembles in many respects the response of invertebrate neurons and cultured mammalian ganglion cells which usually depolarize (Chalazonitis 1963; Brown and Berman 1970; Bingmann et al. 1981a, b; Thomas 1976; Thomas and Meech 1982). More recent experiments have shown that the opposite effects of CO_2 on neuronal membranes are obviously attributable to differences in the extracellular buffer capacity in the microenvironment of the cells (Zidek et al. 1978; Bingmann et al. 1981b). This interpretation is supported, moreover, by the finding that neurons located in the outer layer of hippocampal slices depolarize, whereas cells in the inner layers hyperpolarize with rising pCO_2 (Caspers et al. 1986).

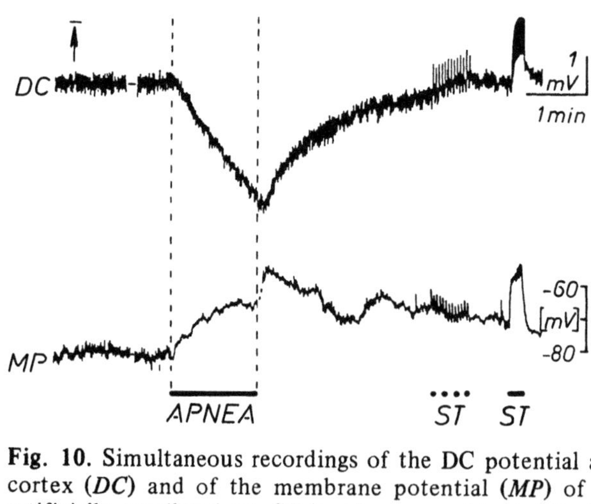

Fig. 10. Simultaneous recordings of the DC potential at the surface of the cerebral cortex (*DC*) and of the membrane potential (*MP*) of a glial cell; anesthetized and artificially ventilated rat. *ST*, low-frequency (*dotted line*) and high-frequency electrical stimulation (*solid line*) of cortical surface. The positive DC shift evoked by rise in the pCO_2 during apnea is accompanied by a depolarization of the glial cell. (From Caspers et al. 1984)

The positive shift of the DC potential evoked by an increase in pCO_2 is accompanied further by changes in the membrane potential of glial cells (Caspers et al. 1984). In contrast to neurons, glial membranes have been found to depolarize with increasing pCO_2. A representative example is displayed in Fig. 10. The depolarizing response of the glial cells is demonstrable throughout the cortex with laminar differences being missing. The behavior of the glial elements during changes in pCO_2 corresponds, moreover, to fluctuations of the $[K^+]_O$; this issue will be dealt with in more detail in the following section.

3.1.3 Correlations with Ionic Activity Changes

3.1.3.1 Changes of $[K^+]_O$

The positive shift of the cortical DC potential evoked by a rise of the pCO_2 is accompanied by an increase in $[K^+]_O$. An example of tracings is displayed in Fig. 11A. Starting from a resting level of 3 mmol/l, the

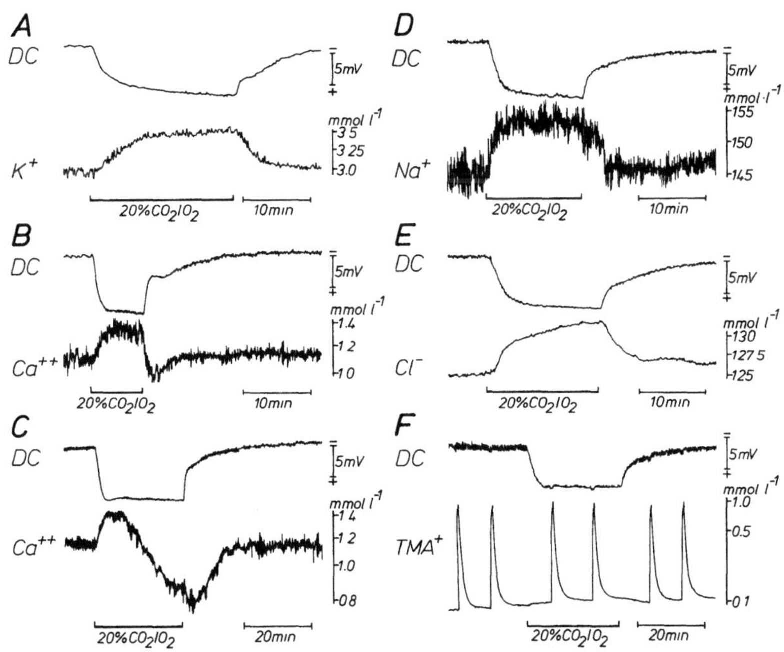

Fig. 11 A–F. Simultaneous recordings for the DC potential, of extracellular ion activities (**A–E**) and of extracellular volume fraction (**F**) during a rise in the CO_2 content of the inspired gas mixture (20% CO_2/O_2, ***horizontal bars***). Somatosensory cortex; anesthetized and artificially ventilated rat. Recordings sites below cortical surface 800 µm (**A**, **B**, **D–F**) and 1000 µm (**C**). In **F**, tetramethylammonium (*TMA^+*) was iontophoretically applied, to estimate the extracellular volume fraction. (**B**, **C**, and **F** from Lehmenkühler et al. 1985)

change of $[K^+]_O$ amounts to 0.25–0.30 mmol/l, when the pCO_2 in the inspired gas mixture is raised by 75 mmHg within a range of 0–300 mmHg (Lehmenkühler et al. 1976; Lehmenkühler 1979; Caspers et al. 1979, 1980, 1984). Despite a certain similarity of the time course in all cortical layers, the bioelectrical and ionic signals differ in several aspects. First, the rise and decay of the evoked DC shift tend to be steeper than in the case of $[K^+]_O$ which, in addition, exhibits a longer latency. Second, the increase in $[K^+]_O$ evoked, for instance, by gradually raising the inspiratory pCO_2, sometimes reaches a maximum at approximately 4 mmol/l, when the concomitant positive DC displacement is still growing in amplitude. Finally, a preceding elevation of the baseline level of $[K^+]_O$, as it occurs, for example, with repetitive seizure attacks can change the CO_2 effect on potassium activity without affecting the DC response (cf. Lehmenkühler and Janus 1978). As a whole, these observations indicate that the positive shift of the cortical DC potential induced by a rise of the pCO_2 is not correlated with the concomitant variations of $[K^+]_O$.

3.1.3.2 Changes of $[Ca^{2+}]_O$

A rise of the pCO_2 in nervous tissue is usually associated with changes of $[Ca^{2+}]_O$. This finding was obtained in studies on rats (Caspers et al. 1979; Lehmenkühler et al. 1984) as well as on cats (Somjen et al. 1986). In the rat cortex a gradual increase of the CO_2 content in the inspired gas mixture from 0% to 20% enhances $[Ca^{2+}]_O$ from the baseline level of 1.1 mmol/l to maximum value of 1.4 mmol/l throughout the cortex (Fig. 11B). Comparing the changes of calcium activity to the concomitant shifts of the DC potential, it becomes evident that the responses of the two signals vary in some essential aspects. First, there are differences in the slope of rise and decay. Second, upon CO_2 withdrawal $[Ca^{2+}]_O$ shows a transient undershoot prior to its return to the baseline, with no corresponding wave being detectable in the DC recording (Fig. 11B). Third, with long-lasting administrations of CO_2 the initial increase of $[Ca^{2+}]_O$ changes to a progressive decline going far below the starting level, while the shift of the DC potential is still running monophasically positive (Fig. 11C).

3.1.3.3 Changes of $[Na^+]_O$ *and* $[Cl^-]_O$

The displacement of the cortical DC potential during hypercapnia is associated with distinct changes of $[Na^+]_O$ and $[Cl^-]_O$ in cortical tissue. Whereas earlier measurements at the surface of the cerebral cortex and in the cerebrospinal fluid yielded controversial results, studies by means of liquid-membrane ion-selective microelectrodes demonstrate a clear-cut monophasic rise of $[Na^+]_O$ when the pCO_2 is raised. An example of tracings is displayed in Fig. 11D, showing an increase of $[Na^+]_O$ by 10 mmol/l above the resting level of 155 mmol/l at an inspiratory CO_2 content of 20%. Corrected for the Ca^{2+} error of the Na^+ selective membrane, the actual increase amounts to 9 mmol/l. The rates of rise and decay of the Na^+ signal are reflected in the concomitant shift of the cortical DC potential.

Measurements with ion-selective microelectrodes based on Corning's chloride exchanger showed $[Cl^-]_O$ to increase during hypercapnia (Fig. 11E). With slight enhancements of the pCO_2 elicited, for instance, by administration of 3%–5% CO_2 in the inspired gas mixture, the dominant elevation of $[Cl^-]_O$ is usually preceded by a moderate fall. Taking into account an HCO^-_3 interference of the Cl^- exchanger, the actual rise of $[Cl^-]_O$ evoked by a gas mixture containing 20% CO_2 ranges from 4 to 5 mmol/l above the normal level of 125 mmol/l. After CO_2 withdrawal the changes of the bioelectrical and ionic signals return to baseline at different times. As a whole, the shifts of the cortical DC potential during hypercapnia are related only in part to the concomitant variations of $[Na^+]_O$ and $[Cl^-]_O$.

3.1.3.4 *Changes of* $[H^+]_O$

At normal values of the arterial pCO_2 the extracellular pH of the brain tissue was found to be independent of the recording depth and to range between 7.2 and 7.35 (Cragg et al. 1977; Lehmenkühler et al. 1981, 1986; Kraig et al. 1983; Nicholson et al. 1985; Mutch and Hansen 1984, 1985; Siesjö et al. 1985). In the rat cortex a rise of the inspiratory pCO_2 usually causes a lowering of the pH, the time course of which differs in some aspects from that of the positive DC deviation. An example of tracings is displayed in Fig. 12A. After CO_2 withdrawal the primary acidic reaction of cortical tissue changes to a transient alkaline shift followed by a secondary acidic wave, before the baseline level of the pH is reached again. Constant differences between various cortical layers are not discernible from in depth profiles of the bioelectrical and ionic responses. Occasionally, another type of pH shift can be observed during hypercapnia (Fig. 12B). It consists of a transient acidic reaction at the beginning of CO_2 administration and of a pronounced alkaline wave after CO_2 has been removed. Apart from a reduced amplitude, the positive shift of the DC potential remains unaffected in these cases. In summary, the DC deviations evoked by a rise of the pCO_2 are not related to the concomitant changes of the extracellular pH.

In order to estimate the contribution of a transmembranous water flux to the changes of ion activities occurring with hypercapnia, the

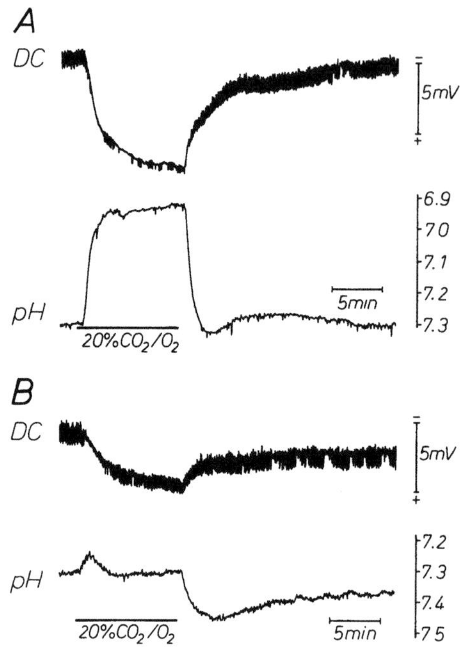

Fig. 12A, B. Simultaneous recordings of the cortical DC potential and of extracellular pH during rise of the inspiratory CO_2 content (***horizontal bars***). **A, B** Examples of different responses in two experiments; anesthetized and artificially ventilated rat. Double-barrelled ion-selective liquid membrane electrode; recording site ca. 800 μm below cortical surface. (Experiments in collaboration with U. Kersting)

extracellular volume fraction has been determined by means of the TMA^+ iontophoresis technique before, during and after CO_2 administration (Lehmenkühler et al. 1985); a typical finding is presented in Fig. 11F. The results indicate that the extracellular volume fraction is not significantly altered by CO_2 exposure.

3.2 Changes of pO_2

3.2.1 Hypoxia and Primary Asphyxiation

In anesthetized animals artificially ventilated at a constant rate, a reduction of the oxygen content in the inspired gas mixture has been found to evoke negative DC displacements in surface leads (Caspers 1963; Caspers et al. 1963; Hirsch et al. 1968a). In a medium range, the amplitude of the DC shift is approximately linearly related to the decrease of the inspiratory pO_2. With further progression of oxygen deficiency, however, the relative enhancement of the negative DC response is strongly reduced and finally followed by a positive deflection. In anesthetized animals, such diphasic effects of hypoxia are usually accompanied by a lowering of heart frequency as well as by a fall in arterial blood pressure. A representative sample of tracings is displayed in Fig. 13A. The critical

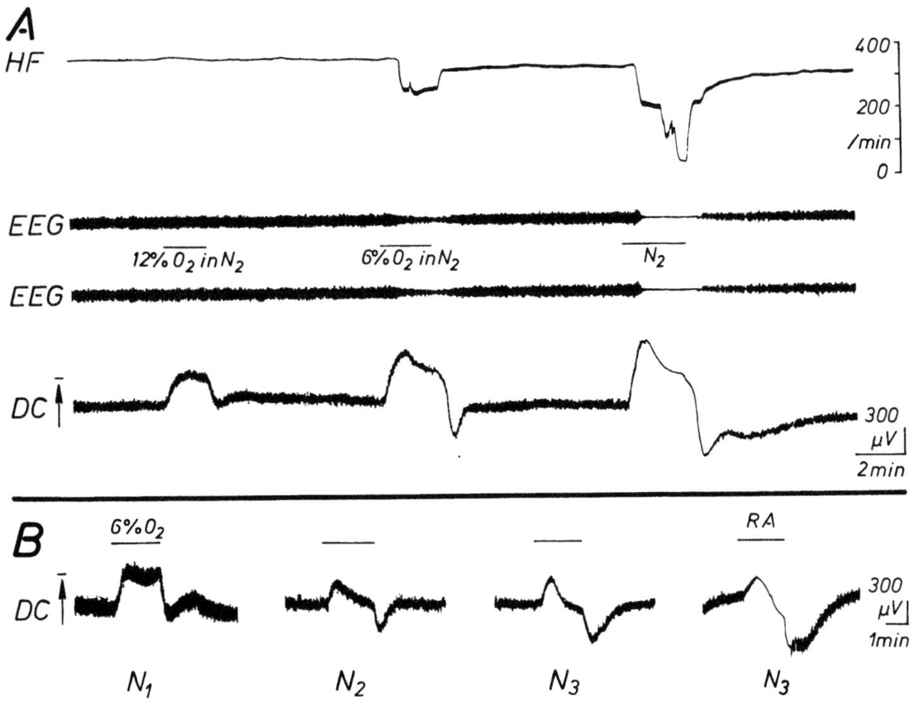

Fig. 13A, B. Cortical DC shifts evoked by lowering of the oxygen content in the inspiratory gas mixture and by respiratory arrest (*RA*); anesthetized and artificially ventilated rat. **A** Administration of 12% O_2 in N_2, 6% O_2 in N_2, and pure N_2. The conventional EEG of both hemispheres and the heart frequency (*HF*) were simultaneously recorded. **B** Application of 6% O_2 in N_2 (N_1–N_3) *and RA*. N_1' DC response in light anesthesia; N_2, N_3, DC response to the application of the same gas mixture (*horizontal bars*) with increasing depth of barbiturate anesthesia. The responses in N_3 resemble each other. (**A** from Caspers et al. 1963; **B** from Caspers and Speckmann 1974)

inspiratory pO_2, at which such a reversal of the DC shift takes place, depends on different mechanisms and may therefore vary within a wide range. In this context the depth of anesthesia plays a decisive role. As shown in Fig. 13B, the administration of 6% in N_2 evokes a monophasic negative DC displacement in the slightly anesthetized animal, whereas the same gas mixture elicits diphasic DC reactions with the initial negative shift increasingly being reduced, when the depth of anesthesia is progressively raised. It has been shown that such complex changes of the DC potential during hypoxia always coincide with an impairment of circulation and/or of respiration giving rise to an increase in pCO_2 and to a lowering of the pH. From these observations the conclusion has been drawn that the proper response of the cortical DC potential to a reduction of tissue pO_2 consists in the negative shift (Caspers and Speckmann

1974). Complex negative-positive or even polyphasic DC reactions appearing during progressive hypoxia are obviously attributable to interferences of oxygen deficiency, hypercapnia, and acidosis, as also occur with primary asphyxiation. Such an interpretation is substantiated, moreover, by the results of unit recordings which will be dealt with in a subsequent section.

Complex DC reactions have been found to occur with primary asphyxiation initiated by respiratory or circulatory arrest(Leao 1947; Goldring et al. 1953, van Harreveld and Stamm 1953a, b; Kempinsky 1954; Meyer and Denny-Brown 1955; Bureš and Burešová 1957; Caspers et al. 1963; Caspers and Speckmann 1971; O'Leary and Goldring 1964; Eidelberg et al. 1967; Speckmann and Caspers 1967; Hirsch et al. 1968c). Typical examples of DC shifts following respiratory arrest are illustrated in Fig. 14. The tracings indicate that the first response of the DC component during asphyxiation consists in a low-voltage negative deviation that turns to an intermediate positive shift within 30–50 s after the onset of respiratory arrest. The transition to the positive displacement is usually characterized by the disappearance of the conventional EEG.

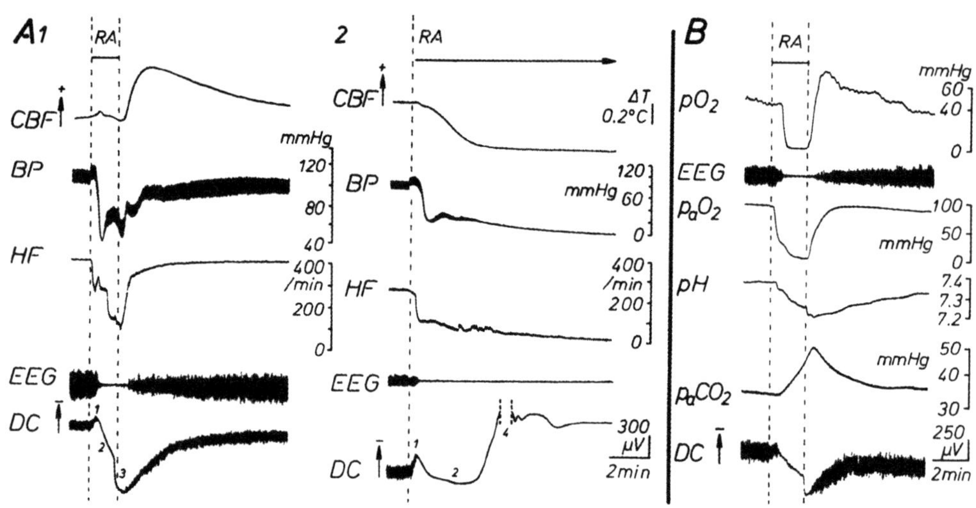

Fig. 14A. B. Relations between cortical DC shifts and changes of circulatory parameters (**A**) and gas pressures (**B**) during respiratory arrest (*RA*) of various durations (*dashed vertical lines* in the anesthetized and artificially ventilated rat. **A** DC deviations and associated changes of conventional *EEG*, heart frequency (*HF*), arterial blood pressure (*BP*), and cortical blood flow (*CBF*). The typical phases of the DC response are indicated by ***numbers 1–4*** (*1*, initial negativity; *2*, intermediate positivity; *3*, reactive positivity; *4*, terminal negativity). The tracings in *1* were obtained during the first, the tracings in *2* after repeated asphyxiations. **B** DC shifts and cortical pO_2, arterial pO_2 (p_aO_2), arterial pH, arterial pCO_2 (p_aCO_2), and the conventional EEG. (From Caspers and Speckmann 1971)

If ventilation is reestablished within the so-called revival time of the animal, the intermediate positive DC shift passes to an additional and steep positive deflection which has been labeled as reactive positivity (Speckmann and Caspers 1967). In case of persisting asphyxiation, the intermediate positive DC displacement proceeds to a high-voltage negative deflection known as "anoxic terminal negative" shift. Probably due to its prominent amplitude, the terminal negative shift has often been described as the only significant DC deviation to occur after respiratory or circulatory arrest, whereas the sequence of potential fluctuations preceding this event has been mentioned or subjected to more detailed studies only in a few papers.

Simultaneous recordings of the cortical DC potential and gas pressures in blood and tissue confirmed that the various components of the complex DC response to primary asphyxiation can be attributed to actual changes in pO_2, pCO_2, and pH. The relationships between these phenomena in a single experiment are displayed in Fig. 14B. As a whole, the findings showed that both the initial and terminal negative DC shifts after respiratory or circulatory arrest depend on the development of tissue anoxia. Thus, both components are lacking during apnea. As described above, with this type of respiratory arrest, tissue pO_2 may range above normal levels for a rather long time, while the pCO_2 increases progressively. In such experimental situations only monophasically positive DC displacements have been observed (cf. Figs. 7B, 8C, and 9A). The findings suggest, furthermore, that both the intermediate and reactive positive DC deflections occurring during primary asphyxiation can be attributed to a rise in pCO_2 and/or to a lowering of the pH. This interpretation is supported by a number of additional findings. With repetitive asphyxiations in shorter intervals the mean level of the pCO_2 tends to increase, while the pH shows a corresponding fall. At the same time the phasic changes in pCO_2 and in pH evoked by each single respiratory arrest are progressively reduced. These effects are clearly reflected in the DC recordings. First, the general rise of the pCO_2 and the lowering of the pH coincide with a continuous positive shift of the DC baseline. Second, the diminution of the phasic responses of the pCO_2 and pH is accompanied by a decrease of the intermediate as well as of the reactive positive DC deflections (Speckmann and Caspers 1967). Finally, the intermediate positive DC displacement may be reduced to an extent to compensate for the initial negative shift. Besides technical reasons these findings may account for the fact that the DC deviations preceding the final negative shift have frequently been overlooked. As a whole the results indicate that negative DC deviations which appear in the course of primary asphyxiation can be attributed to the development of anoxia, whereas the positive components interspersed between the negative shifts depend on the simultaneous changes in the pCO_2 and/or pH. This interpretation is substantiated moreover by findings on single neurons which will be dealt with in a subsequent section.

The results of DC recordings during primary asphyxiation demonstrate that considerable DC shifts are still occurring at a time when the waves of the conventional EEG (in the course of circulatory and/or respiratory arrest) have been completely extinguished. These observations suggest that DC recordings might yield additional information with regard to possibilities of reviving the brain. Consequently, a number of studies have been devoted particularly to this problem (Speckmann and Caspers 1967; Hirsch et al. 1968c; Caspers and Speckmann 1971). It has been shown that DC recordings, to some extent, actually allow more precise prediction of the reactivation of cortical activity than does the EEG. Thus, reventilation proves effective in restoring normal cortical functions in each case where it is resumed during primary or intermediate DC deflections. In the course of the terminal negative shift, however, reventilation tends increasingly to fail. Finally, the probability of revival becomes zero as soon as a terminal negative shift has reached its maximal value. These relations have also been shown to exist with considerable fluctuations of the latency of the terminal DC deflections, which may vary between 1 and approximately 8 min. Because the waves of the conventional EEG disappear within 30–50 s after the onset of asphyxiation, predictions comparable to those derived from DC recordings cannot be made from the typical AC registration of cortical activity.

The experiences described above suggested that the terminal negative shift of the cortical DC potential during asphyxiation might represent a suitable criterion for the so-called revival time of the brain (cf. Schneider 1965). However, such an assumption has been disproved by further investigations (Caspers and Speckmann 1971). Indeed, it has been shown that the described relations between DC displacements and probability of revival are valid provided that the procedures of reanimation are restricted to reventilation. In case of some additional measures, such as heart massage, the DC deviations as well as the extinction of the conventional EEG are often found to be completely reversible even at a time when the terminal negative shift has passed its maximal value. These observations suggest that the development of the terminal negative DC shift during asphyxiation reflects the increasing insufficiency of circulation, rather than the revival time of the brain tissue. Various experimental procedures have been adopted to further clarify this problem. Hirsch et al. (1968a) used isolated brains perfused by a separate donator. Speckmann and Caspers (1966), on the other hand, performed similar studies by means of a heart-lung machine or by applying gas mixtures of low O_2 contents after clamping both common carotid arteries. These experiments showed that the extinction of the conventional EEG is completely reversible as long as the arrest of respiration and circulation does not exceed a mean duration of 8–10 min. If asphyxiation persists beyond this period the chance of EEG recovery is progressively reduced, until each signal of reactivation fails, whenever ventilation or circulation are resumed after 30 min. These findings correspond rather closely to earlier observations by Hirsch (1957). Later observations reported by Hossmann (1982), on the existence of essentially longer revival times of the brain tissue, are not immediately comparable to the described DC findings in rats because of the different techniques applied.

3.2.1.1 Correlations with Neuronal and Glial Activity

As compared to the actions of CO_2 on neuronal activity, the effects of oxygen deficiency on single neuronal elements proved more uniform, provided that the lowering of the tissue pO_2 remained uncomplicated by essential secondary changes in pCO_2 and pH (cf. Kolmodin and Skoglund 1959; Nelson and Frank 1963; Collewijn and van Harreveld 1966; Eccles et al. 1966; Glötzner 1967; Niechaj and van Harreveld 1968; Speckmann et al. 1970; Caspers and Speckmann 1971, 1972, 1974; Speckmann and Caspers 1974; Hansen 1985). As a rule, negative shifts of the cortical DC potential evoked by a stepwise reduction of the pO_2 have been found to correspond fairly closely to a characteristic neuronal reaction consisting of a decrease in membrane potential, an increase in amplitude and frequency of occurrence of depolarizing postsynaptic potentials, and of a rise of discharge frequency (Caspers and Speckmann 1974).

The rather simple correlations between DC shifts and neuronal activity changes found with states of pure oxygen deficiency become more complex if hypoxia progresses. Under this condition the lowering of pO_2 causes impairments of circulation and thus changes in pCO_2 and pH. Such an experimental situation rapidly develops, for instance, with breathing pure nitrogen or with interruption of artificial ventilation. In these cases the effects of hypoxia on neuronal activity interfere with those exerted by changes in pCO_2 and pH. As pointed out in an earlier section, these complex effects are different depending on whether I or E neurons are involved. In I neurons, which largely predominate in cerebral structures, primary asphyxiation evokes an initial increase in discharge frequency followed by a rapid decay. These variations of the

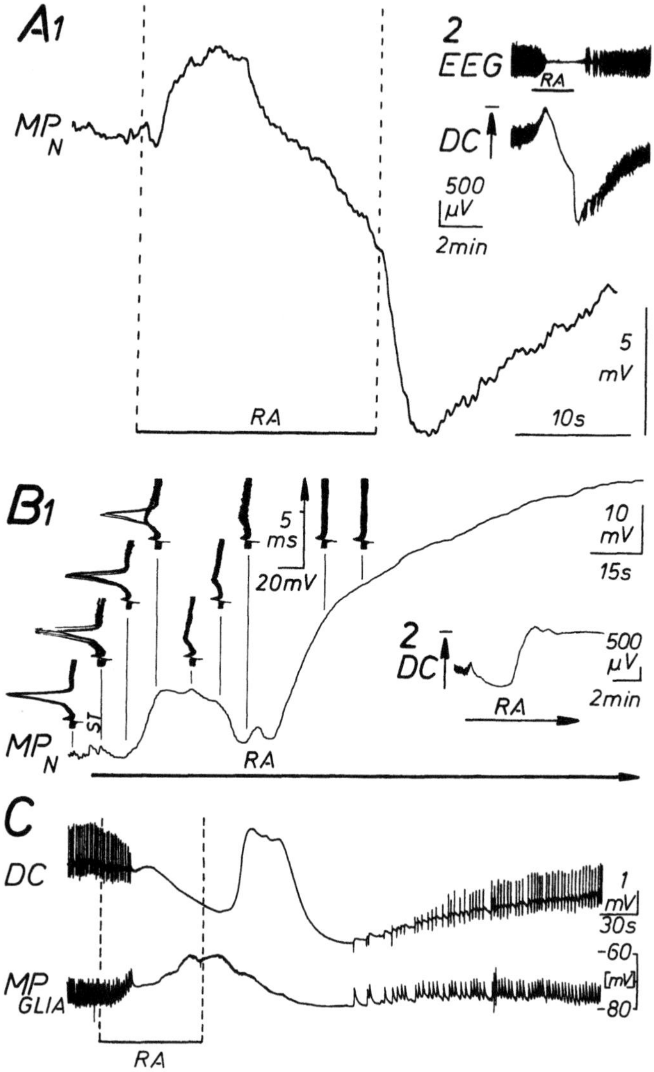

Fig. 15 A–C. Relations between cortical DC potential and membrane potential changes (*MP*) of neurons (**A**, **B**) and glial cells (**C**) during respiratory arrest (*RA*); anesthetized and artificially ventilated rat. **A** Transient RA (*dashed vertical lines*). Changes of neuronal MP (MP_N) during RA (**A1**) and of *DC* potential and conventional *EEG* in the same experimental situation (*A2*). **B** Persistent RA. Changes of MP of a spinal motoneuron (MP_N) during *RA* (*B1*) and of *DC* potential in the same experimental situation (*B2*). Excitatory postsynaptic potentials and action potentials evoked by supramaximal dorsal root stimulation (*ST*) were simultaneously recorded. In each single tracing five sweeps are superimposed. MP shifts and postsynaptic responses are related to each other by *vertical lines*. **C** Simultaneous recordings of DC potential and of MP of a cortical glial cell (MP_{GLIA}) during *RA* (*dashed vertical lines*). (**A** and **B** from Caspers and Speckmann 1971)

firing rate run parallel to an early depolarization and a subsequent re- and hyperpolarization of the cell. In their turn, these changes in membrane potential correspond to the sequence of negative-positive DC deflections described in the preceding section. An example of such a recording is presented in Fig. 15A. Since this experiment was carried out using double microelectrodes, a distortion of the MP tracing by picking up extracellular DC potentials via a remote reference electrode can be neglected. On this condition the shifts of the membrane potential illustrated in Fig. 15A are found to occur in all I neurons independent of whether they are located in cortical, subcortical, or spinal structures. Furthermore, comparative measurements showed that the various phases of the MP shifts are statistically synchronized with DC deflections recorded using gross electrodes in the same area. In single experiments the coincidence between the two bioelectrical events may vary to some extent. Such differences can be attributed to the fact that the development of anoxia and hypercapnia after the onset of asphyxiation shows local variations probably due to fluctuations in local blood flow. The accordance between DC displacement and MP shifts illustrated in Fig. 15A can be observed also with irreversible respiratory and/or circulatory arrest. Apart from possible differences in time course, as mentioned above, the terminal negative DC shift at the cortical surface corresponds to a complete depolarization of I neurons both in cerebral and in spinal structures. An example of such a finding is presented in Fig. 15B. The tracings indicate, moreover, that postsynaptic responses to a presynaptic volley disappear already in the course of the initial depolarization of the neuron. This observation corresponds to the early suppression of EEG waves within the peak region of the first negative DC deflection found in cortical leads. Figure 15B shows that the appearance of excitatory postsynaptic potentials (EPSP), evoked by suprathreshold presynaptic stimulation, outlasts the extinction of somatic action potentials after the onset of asphyxiation by up to 20 s. This finding may explain the results of earlier studies which showed that unit discharges in extracellular recordings tend to disappear prior to the waves of the conventional EEG in the course of asphyxiation (Li and Jasper 1953; Creutzfeldt et al. 1957; Baumgartner et al. 1963).

The correspondence between DC shifts and MP changes of I neurons during respiratory arrest is maintained, moreover, on the special conditions of apnea, described in a preceding section. As has already been pointed out, the evoked positive DC deviation runs parallel to a unidirectional hyperpolarization of neurons in this experimental situation. The observation proves, on the one hand, that the initial and terminal depolarizations of the neuron during primary asphyxiation are due to the development of O_2 deficiency in the same way as the early and late surface negative DC shifts. On the other hand, both the increase in membrane potential and the positive DC deflection found in intermediate stages of asphyxiation can be attributed preferentially to a rise in pCO_2 and/or to a lowering in pH.

As compared with the polyphasic reactions of cortical neurons, the responses of glial cells to a progressive lowering in pO_2 are obviously more uniform. In accordance with observations made by Godfraind et al. (1971) our own experiments showed that glial elements depolarize increasingly in all stages of an ongoing O_2 deficiency, i.e., no intermediate hyperpolarization takes place (Fig. 15C). This finding can be explained by a unidirectional rise of extracellular potassium concentration due to both the hypoxia and the secondarily developing hypercapnia and acidosis.

3.2.1.2 Relations with Ionic Activity Changes

The shifts of the cortical DC potential during asphyxiation have been shown to consist of two distinct phases. The first comprises the initial negative wave followed by the intermediate positive deflection. A successful reventilation in this stage, on the one hand, evokes the reactive positive DC displacement. With persisting asphyxiation, on the other hand, the second phase develops characterized by the high-voltage negative DC shift, also known as "anoxic terminal depolarization" (Bureš and Burešová 1957).

Simultaneous measurements of ion activities by means of ion-selective microelectrodes show that the first phase of asphyxiation is associated with a slow increase of all ions determined and with a modest decrease of the extracellular space in the range of 15%. As far as K^+, Ca^{2+}, and H^+ are concerned, the increase in activity exhibits two distinct steps, whereas Na^+ and Cl^- rise monophasically (Fig. 16). The reactive positive DC shift indicating a successful reventilation is accompanied by a steep decrease of $[K^+]_O$ and $[Ca^{2+}]_O$, both showing undershoots before returning to baseline (Staschen et al. 1981). In contrast, the rise of $[Na^+]_O$, $[Cl^-]_O$, and $[H^+]_O$ requires several minutes to recover. As a whole, no dinstinct correlations between DC shifts and changes of ionic activities are detectable in the initial phase of asphyxiation.

The majority of studies dealing with ion activities in states of hypoxia, ischemia, etc. were performed under experimental conditions including the terminal negative DC displacement (cf. Vyskocil et al. 1972; Morris 1974; Dora and Zeuthen 1976; Kirshner et al. 1976; Hansen and Olsen 1980; Hansen and Zeuthen 1981; Harris et al. 1981; Lehmenkühler et al. 1981; Lehmenkühler 1982; Kraig et al. 1983; Mutch and Hansen 1984,

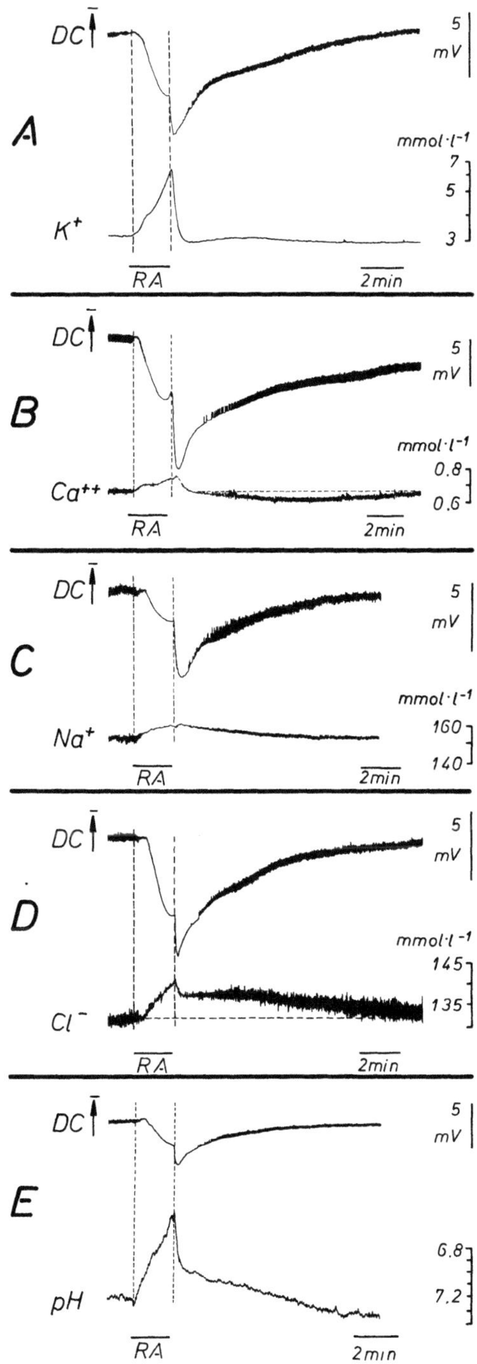

Fig. 16 A–E. Simultaneous recordings of cortical DC potential and of extracellular activity of potassium (K^+), calcium (Ca^{2+}), sodium (Na^+), chloride (Cl^-), and hydrogen (pH) during respiratory arrest (***RA, dashed vertical lines***); anesthetized and artificially ventilated rat. (Experiments in collaboration with M. Staschen)

1985; Hansen 1985). The findings demonstrate that the second phase of the DC deviations during asphyxiation is marked by rapid and large ionic shifts as well as by a dramatic reduction of the extracellular volume by ca. 60%. The time course of the negative DC deflection is correlated only with the change in $[K^+]_O$ that rises up to 60 mmol/l. At the highest slope of the negative DC displacement, corresponding to a $[K^+]_O$ of 12–15 mmol/l, steep sustained decreases of $[Ca^{2+}]_O$ by one decade and of $[Na^+]_O$ and $[Cl^-]_O$ by 50% of the normal level have been described (Nicholson and Kraig 1981; Hansen and Zeuthen 1981; Harris et al. 1981; Hansen 1985). Contrasting to these ionic alterations, the extracellular pH shows a transient alkaline wave superimposed on the slow acidic response that starts immediately in the beginning of the first phase (Lehmenkühler et al. 1981; Lehmenkühler 1982; Kraig et al. 1983; Mutch and Hansen 1984, 1985). These changes of ionic activities and of the extracellular volume fraction resemble those occurring during spreading depression (cf. Nicholson 1981; Hansen 1985).

3.2.2 Hyperoxia

Increases in oxygen pressure above normal levels have also been described to exert a distinct influence on cortical DC potentials (Speckmann and Caspers 1969; Caspers and Speckmann 1974). As can be estimated from the tracings in Fig. 17, the first DC responses to a rise in pO_2 appear at pressure levels as low as 300 mmHg. In animals ventilated artificially at a constant rate, the first DC reaction usually consists in a positive shift that increases in amplitude, if the pO_2 is further elevated. The records in Fig. 17 additionally show that the primary positive deflection may turn to a slowly developing negative shift which becomes more prominent with rising O_2 pressures. The occurrence of such a negative response is clearly reflected by the negative overshooting of the DC potential following the transition to breathing normal room air.

The occurrence of the positive DC deflections has been attributed to simultaneous changes in pCO_2 due to a reduction of the CO_2 binding sites in the venous blood. This interpretation is actually supported by a number of experimental observations. At first, continuous measurements in the carotid arteries and the jugular veins confirmed that each increase of the inspiratory pO_2 is associated with an increase of blood pCO_2 provided that the rate of artificial ventilation remains constant (Fig. 18B). In agreement with this finding cortical blood flow is usually found to rise. Furthermore, it has been shown that the amplitudes of the positive DC shift induced by a definite increase of the inspiratory pO_2 depend on the preexisting CO_2 pressure level. This statement is substantiated by Fig. 18A. Corresponding to the shape of the CO_2 binding

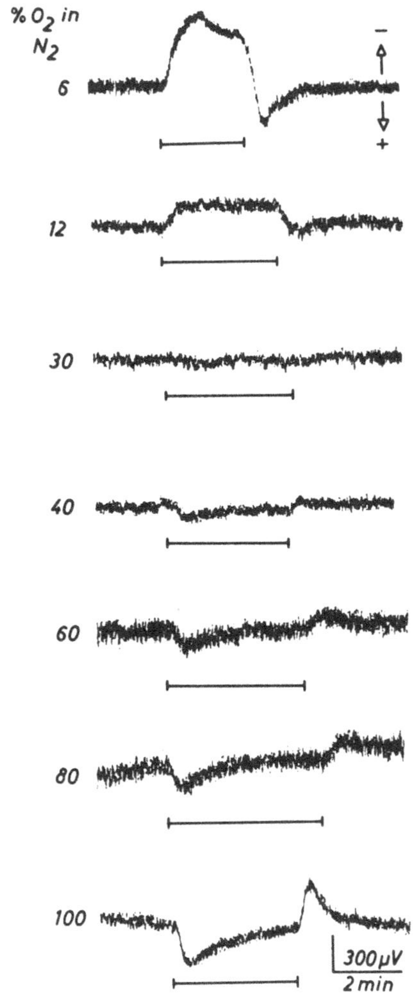

Fig. 17. Cortical DC shifts induced by administration of gas mixtures with increasing oxygen content (6%–100% O_2 ; *left margin*) in the anesthetized and artificially ventilated rat. Application time of either gas mixture is indicated by *horizontal bars* beneath each individual curve.
(From Caspers and Speckmann 1974)

curve, the administration of 100% O_2 gives rise to a distinct, positive DC deflection during artificial hypoventilation, whereas such a response is strongly reduced or even lacking at higher ventilation rates.

In the experiment illustrated in Figs. 17 and 18 the slowly developing negative shift of the DC potential already commenced at inspiratory O_2 contents of 40%–60%. Comparative measurements showed that the critical O_2 value at which this type of DC response takes place may vary within a wide range. The decisive factor seems to consist in the actual rise of tissue pO_2 determined by a number of additional mechanisms, such as gross cortical blood flow, changes in local microcirculation, and in O_2 consumption. This conclusion can be derived from recordings of the DC component and of tissue pO_2 in various experimental

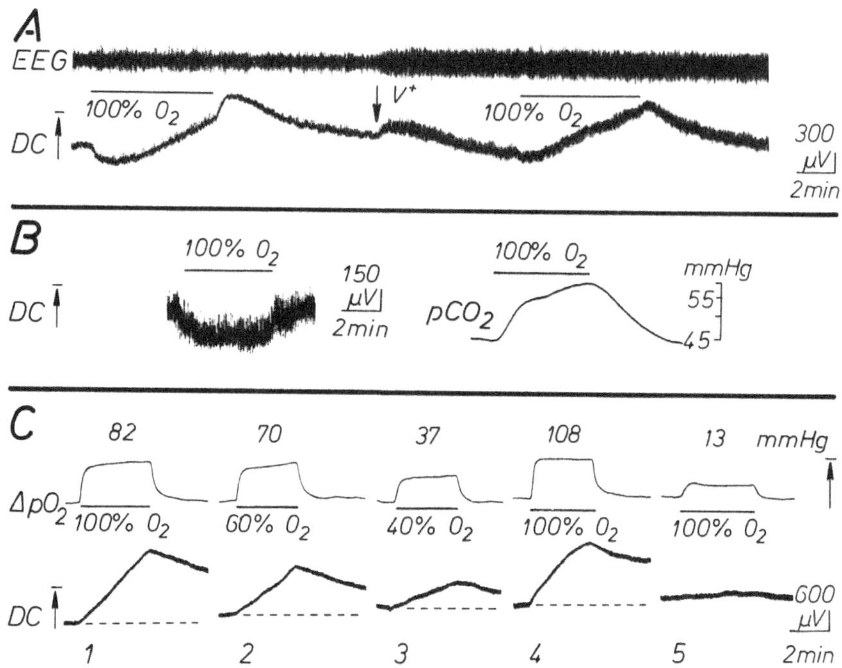

Fig. 18 A–C. Positive-negative cortical DC shifts induced by higher inspiratory oxygen contents in anesthetized rats ventilated artificially at a constant rate. **A** Responses of the cortical DC component and of the conventional EEG to application of 100% oxygen in the inspired gas mixture during normoventilation and hyperventilation ($\dot{V}^{+}$) **B** Positive DC shift and corresponding rise of the arterial pCO_2 during application of 100% oxygen. **C** Relations between negative cortical DC shifts and rise of cortical pO_2 evoked by applying gas mixtures of various O_2 contents. In *C5* arterial blood pressure had been reduced to about 60 mmHg. (From Caspers and Speckmann 1974)

situations. As the tracings in Fig. 18C indicate, the steepness of the evoked negative DC shift is positively related to the actual elevation of cortical pO_2. In additional experiments performed on rats at hyperbaric O_2 pressure levels, the development of negative DC deflections showed further accelerations and reached amplitudes of up to 10 mV. With the critical rise of the inspiratory pO_2 above 2–3 atm, the described deviations of the DC baseline finally merge into complex deflections associated with the onset of seizures (Lehmenkühler et al. 1978). It has therefore been suspected that the slow negative DC displacement preceding this event might reflect an early stage of O_2 intoxication of cortical tissue. However, as yet this problem is not settled.

References

Ayala GF, Walker AE (1965) Steady potential fields in cortical epileptogenesis. Electroencephalogr Clin Neurophysiol 18:519

Ayala GF, Matsumoto H, Gumnit RJ (1970) Excitability changes and inhibitory mechanisms in neocortical neurons during seizures. J Neurophysiol 33:73–85

Ayala GF, Dichter M, Gumnit RJ, Matsumoto H, Spencer WA (1973) Genesis of epileptic interictal spikes. New knowledge of cortical feedback systems suggests a neurophysiological explanation of brief paroxysms. Brain Res 52:1–17

Baumgartner G, Creutzfeldt O, Jung R (1963) Microphysiology of cortical neurons in acute anoxia and in retinal ischaemia. In: Gastaut H, Meyer JS (eds) Cerebral anoxia and the EEG. Thomas, Springfield, pp 5–34

Besson JM, Aleonard P (1965) Variations du potentiel cortical continu sous l'influence de modifications ventilatoires. J Physiol (Paris) 57:558–559

Besson JM, Woody CD, Aleonard P, Thompson HK, Albe-Fessard D, Marshall WH (1970) Correlations of brain d–c shifts with changes in cerebral blood flow. Am J Physiol 218:284–291

Bingmann D, Kolde G, Speckmann E-J (1981a) Reactions of CA3 neurons in hippocampal slices to changes of PCO_2 and pH in the bath solution. Pflügers Arch 391:R32

Bingmann D, Kienecker EW, Caspers H, Knoche H (1981b) Chemoreceptor activity of sinus nerve fibres after their implanation into the wall of the external artery. In: Belmonte C, Pallot DJ, Acker H, Fidone S (eds) Arterial chemoreceptors. Leicester University Press, Leicester, pp 92–101

Bishop GH, O'Leary JL (1950) The effect of polarizing currents on cell potentials and other significance in the interpretation of central nervous activity. Electronencephalogr Clin Neurphysiol 2:401–416

Brazier MAB (1959) The historical development of neurophysiology. In: Field J (ed) Neurophysiology, vol 1. American Physiological Society, Washington, pp 1–58 (Handbook of physiology, sect 1)

Brazier MAB (1963) The discoverers of the steady potentials of the brain: Caton and Beck. UCLA Forum Sci 1:1–14

Brown AM, Berman PR (1970) Mechanism of excitation of Aplysia neurons by carbon dioxide. J Gen Physiol 56:543–558

Bureš J, Burešova O, (1957) Die anoxische Terminaldepolarisation als Indicator der Vulnerabilität der Großhirnrinde bei Anoxie und Ischämie. Pflügers Arch 264: 325–334

Bureš J, Burešova O, Krivanek J (1974) The mechanism and applications of Leao's spreading depression of electroencephalographic activity. Academic, New York

Carpenter DO, Hubbard JH, Humphrey DR, Thompson HK, Marshall WH (1974) Carbon dioxide effects on nerve cell function. In: Nahas G, Schaefer KE (eds) Carbon dioxide and metabolic regulations. Springer, Heidelberg New York Berlin, pp 49–62 (Topics in environmental physiology and medicine)

Caspers H (1959) Über die Beziehungen zwischen Dendritenpotential und Gleichspannung an der Hirnrinde. Pflügers Arch 269:157–181

Caspers H (1963) Relations of steady potential shifts in the cortex to the wakefulness sleep spectrum. In: Brazier MAB (ed) Brain function, University of California Press, Berkeley, pp 177–200

Caspers H (1965) Shifts of the cortical steady potential during various stages of sleep. In: Jouvet M (ed) Aspects anatomo-fonctionnels de la physiologie du sommeil. Centre National de la Recherche Scientifique, Paris, 213–229

Caspers H (1974) Preface. In: A Remond (ed) Handbook of electroencephalography and clinical neurophysiology, vol 10/A. Elsevier, Amsterdam, p 10A/3

Caspers H, Speckmann E-J (1969) DC potential shifts in paroxysmal states. In: Jasper HH, Ward AA, Pope A (eds) Basic mechanisms of the epilepsies. Little Brown, Boston, pp 375–388

Caspers H, Speckmann E-J (1971) Gleichspannungsverschiebungen an der Hirnrinde bei Asphyxie. Ärztl Forsch 25:241–255

Caspers H, Speckmann E-J (1972) Cerebral pO_2, pCO_2 and pH: Changes during convulsive activity and their significance for spontaneous arrest of seizures. Epilepsia (Amst) 13:699–725

Caspers H, Speckmann E-J (1974) Cortical DC shifts associated with changes of gas tensions in blood and tissue. In: Remond A (ed) Handbook of electroencephalography and clinical neurophysiology, vol 10 A. Elsevier, Amsterdam, pp 10A/41–10A/65

Caspers H, Schütz E, Speckmann E-J (1963) Gleichspannungsänderungen an der Hirnrinde bei Sauerstoffmangel. Z Biol 114:112–126

Caspers H, Speckmann E-J, Lehmenkühler A (1979) Effects of CO_2 on cortical field potentials in relation to neuronal activity. In: Speckmann E-J, Caspers H (eds) Origin of cerebral field potentials. Thieme, Stuttgart, pp 151–163

Caspers H, Speckmann E-J, Lehmenkühler A (1980) Electrogenesis of cortical DC potentials. Prog Brain Res 54:3–15

Caspers H, Speckmann E-J, Lehmenkühler A (1984) Electrogenesis of slow potentials of the brain. In: Elbert T, Rockstroh B, Lutzenberger W, Birbaumer N (eds) Self-regulation of the brain and behavior. Springer, Berlin Heidelberg New York, pp 27–41

Caspers H, Speckmann E-J, Bingmann D, Lehmenkühler A (1986) Wirkungen von CO_2 auf das Membranpotential einzelner Neurone. In: Grote J, Thews G (eds) Aktuelle Probleme der Atmungs- und Kreislaufregulation. Steiner, Stuttgart, pp 185–195 (Funktionsanalyse biologischer System, vol 15)

Castelluci VF, Goldring S (1970) Contribution to steady potential shifts of slow depolarization in cells presumed to be glia. Electroencephalogr Clin. Neurophysiol 28:109–118

Caton R (1875) The electrical currents of the brain. Br Med J. 2:278

Chalazonitis N (1963) Effects of changes in pCO_2 and pO_2 on rhythmic potentials from giant neurons. Ann NY Acad Sci 109:451–479

Cohen MW (1974) Glial potentials and their contribution to extracellular recordings. In: Remond A (ed) Handbook of electroencephalography and clinical neurophysiology, vol 2/B. Elsevier, Amsterdam, pp 2B/43–2B/60

Collewijn H, van Harreveld A (1966) Intracellular recording from cat spinal motoneurons during acute asphyxia. J Physiol (Lond) 185:1–14

Cragg P, Patterson L, Purves MJ (1977) The pH of brain extracellular fluid in the cat. J Physiol (Lond) 272:137–166

Creutzfeldt OD, Houchin J (1974) Neuronal basis of EEG waves. In: Remond A (ed) Handbook of electroencephalography and clinical neurophysiology, vol 2C. Elsevier, Amsterdam, pp 2C/5–2C/55

Creutzfeldt OD, Kasamatsu A, Vaz-Ferreira A (1957) Aktivitätsänderungen einzelner corticaler Neurone im akuten Sauerstoffmangel und ihre Beziehungen zum EEG bei Katzen. Pflügers Arch 263:647–667

Davis H, Davis PA, Loomis AL, Harvey EN, Hobart G (1939) A search for changes in direct current potentials of the head during sleep. J Neurophysiol. 2:129–135

Dichter M, Spencer WA (1969a) Penicillin-induced interictal discharge from the cat hippocampus. I. Characteristics and topographical feature. J Neurophysiol 32:649–662

Dichter M, Spencer WA (1969b) Penicillin-induced interictal discharge from the cat hippocampus. II. Mechanism underlying origin and restriction. J Neurophysiol 32:663–687

Dietzel I, Heinemann U, Hofmeier G, Lux HD (1980) Transient changes in the size of the extracellular space in the sensorimotor cortex of cats in relation to stimulus-induced changes in potassium concentration. Exp Brain Res 40:432–439

Dietzel I, Heinemann U, Hofmeier G, Lux HD (1982a) Changes in the extracellular volume in the cerebral cortex of cats in relation to stimulus-induced epileptiform afterdischarges. In: Klee MR, Lux HD, Speckmann E-J (eds) Physiology and pharmacology of epileptogenic phenomena. Raven, New York, pp 5–12

Dietzel I, Heinemann U, Hofmeier G, Lux HD (1982b) Stimulus-induced changes in extracellular Na^+ and Cl^- concentration in relation to changes in the size of the extracellular space. Exp Brain Res 46:73–84

Dora E, Zeuthen T (1976) Brain metabolism and ion movements in the brain cortex of the rat during anoxia. In: Kessler M, Clark LC Jr, Lübbers DW, Silver IA, Simon W (eds) Ion and enzyme electrodes in biology and medicine. Urban and Schwarzenberg, München, pp 294–298

Eccles RM, Løyning Y, Oshima T (1966) Effects of hypoxia on the monosynaptic reflex pathway in the cat spinal cord. J Neurophysiol 29:315–332

Eidelberg E, Meyerson BA (1964) Effects of lidocaine on cortical dendritic activity. Arch int Pharmacodyn. 147:576–858

Eidelberg E, Kolmodin GM, Meyerson BA (1967) Effect of asphyxia on the cortical steady potential in adult and fetal sheep. Acta Physiol Scand 69:257–261

Elger CE, Speckmann E-J (1980) Focal interictal epileptiform discharges (FIED) in the epicortical EEG and their relations to spinal field potentials in the rat. Electroencephalogr Clin Neurophysiol 48:447–460

Elger CE, Speckmann E-J (1983) Penicillin-induced epileptic foci in the motor cortex: vertical inhibition. Electroencephalogr Clin Neurophysiol 56:604–622

Elger CE, Speckmann E-J, Prohaska O, Caspers H (1981) Pattern of intracortical-potential distribution during focal interictal epileptiform discharges (FIED) and its relation to spinal field potentials in the rat. Electroencephalogr Clin Neurophysiol 51:393–402

Ferguson JH, Jasper HH (1971) Laminar DC studies of acetylcholine-activated epileptiform discharges in cerebral cortex. Electroencephalogr Clin Neurophysiol 30:377–390

Glötzner F (1967) Intracelluläre Potentiale, EEG and corticale Gleichspannung an der sensomotorischen Rinde der Katze bei akuter Hypoxie. Arch Psychiat Nervenkr 210:274–296

Godfraind JM, Kawamura H, Krnjevic K, Pumain R (1971) Actions of dinitrophenol and some other metabolic inhibitors on cortical neurons. J Physiol (Lond) 215:199–222

Goldensohn ES, Schoenfeld RL, Hoefer PFA (1951) The slowly changing voltage of the brain and the elctroencephalogram. Electroencephalogr Clin Neurophysiol 3:231–236

Goldring S, O'Leary JL, Lam RL (1953) Effect of malononitril upon the electroencephalogram of the rabbit. Electroencephalogr Clin Neurophysiol 5:395–400

Gumnit R (1974a) Introduction and recording techniques. In: Remond A (ed) Handbook of electroencephalography and clinical neurophysiology, vol 10/A. Elsevier, Amsterdam, pp 10A/5–10A/11

Gumnit R (1974b) DC shifts accompanying seizure activity. In: Remonad A (ed) Handbook of electroencephalography and clinical neurophysiology, vol 10/A. Elsevier, Amsterdam, pp 10A/66–10A/87

Gumnit R (1979) Field potentials in partial seizures – a reappraisal. In: Speckmann E-J, Caspers H (eds) Origin of cerebral field potentials. Thieme, Stuttgart, pp 183–194

Gumnit R, Takahashi T (1965) Changes in direct current activity during experimental focal seizures. Electroencephalogr Clin Neurophysiol 19:63–74

Gumnit R, Matsumoto H, Vasconetto C (1970) DC activity in the depth of an experimental focus. Electroencephalogr Clin Neurophysiol 28:333–339
Haider M, Groll-Knapp E, Ganglberger JA (1981) Event-related slow (DC) potentials in the human brain. Rev Physiol Biochem Pharmacol 88:125–197
Hansen AJ (1985) Effect of anoxia on ion distribution in the brain. Physiol Rev 65: 101–148
Hansen AJ, Olsen CE (1980) Brain extracellular space during spreading depression and ischemia. Acta Physiol Scand 108:355–365
Hansen AJ, Zeuthen T (1981) Extracellular ion concentration during spreading depression and ischemia in the rat brain cortex. Acta Physiol Scand 113:437–445
Harris RJ, Symon L, Branston NM, Bayhan M (1981) Changes in extracellular calcium activity in cerebral ischaemia. J Cereb Blood Flow Metab 1:203–209
Heinemann U, Lux HD (1977) Ceiling of stimulus induced rises in extracellular potassium concentration. Brain Res 120:231–249
Heinemann U, Lux HD, Gutnick J (1977) Extracellular free calcium and potassium during paroxysmal activity in the cerebral cortex of the cat. Exp Brain Res 27:237–243
Heinemann U, Lux HD, Gutnick J (1978) Changes in extracellular free calcium and postassium activity in the somatosensory cortex of cats. In: Chalazonitis N, Boisson M (eds) Abnormal neuronal discharges. Raven, New York, pp 329–345
Heinemann U, Lux HD, Marciani MG, Hofmeier G (1979) Slow potentials in relation to changes in extracellular potassium activity in the cortex of cats. In: Speckmann E-J, Caspers H (eds) Origin of cerebral field potentials. Thieme, Stuttgart, pp 33–48
Heinemann U, Konnerth A, Lux HD (1981) Stimulation induced changes in extracellular free calcium in normal cortex and chronic alumina cream foci of cats. Brain Res 213:246–250
Heinemann U, Konnerth A, Pumain R, Wadman WJ (1986) Extracellular calcium and potassium concentration changes in chronic epileptic brain tissue. Adv Neurol pp 641–661
Hirsch H, Euler KH, Schneider M (1957) Über die Erholung und Wiederbelebung des Gehirns nach Ischämie bei Normothermie. Pflügers Arch 265:281–313
Hirsch H, Scholl H, Dickmans HA, Eisolt J, Gaehtgens P, Mann H, Krankenhagen B (1968a) Die corticale Gleichspannung des Hundegehirns bei Veränderung des arteriellen pO_2 and pCO_2. Pflügers Arch 301:344–350
Hirsch H, Scholl H, Dickmans HA, Eisolt, J, Gaehtgens P, Mann H, Krankenhagen B (1968b) Über die corticale Gleichspannung nach Überschreiten der Wiederbelebungszeit des Gehirns. Pflügers Arch 301:351–357
Hirsch H, Scholl H, Paschke KG, Schmid-Schönbein H (1968c) Die Veränderungen der corticalen Gleichspannung bei kompletter und inkompletter Ischämie des Gehirns. Pflügers Arch 301:334–343
Hossmann K-A (1982) Treatment of experimental cerebral ischemia. J Cereb Blood Flow Metab 2:275–297
Hotson JR, Sypert GW, Ward AA (1973) Extracellular potassium concentration changes during propagated seizures in neocortex. Exp Neurol 38:20–26
Kempinsky WH (1954) Steady potential gradients in experimental cerebral occlusion. Electroencephalogr Clin Neurophysiol 6:375–388
Kirshner HS, Blank WF Jr, Myers RE (1976) Changes in cortical subarachnoid fluid potassium concentrations during hypoxia. Arch Neurol 33:84–90
Köhler W, Held R, O'Connel DW (1982) An investigation of cortical currents. Proc Philos Soc 96:290–330
Kolmodin GM, Skoglund CR (1959) Influence of asphyxia on membrane potential level and action potentials of spinal moto- and interneurons. Acta Physiol Scand 45:1–18

Kraig RP, Ferreira-Filho CR, Nicholson C (1983) Alkaline and acid transients in cerebellar microenvironment. J Neurophysiol 49:831–850

Labeyrie E, Koechlin Y (1979) Photoelectrode with a very short time-constant for recording intracerebrally Ca^{++} transients at a cellular level. J Neurosci Methods 1:35–39

Leao AAP (1944) Spreading depression of activity in the cerebral cortex. J Neurophysiol 7:359–390

Leao AAP (1947) Further observations of the spreading depression of activity in the cerebral cortex. J Neurophysiol 10:409–414

Lehmenkühler A, Speckmann E-J, Caspers H (1976) Cortical spreading depression in relation to potassium activity, oxygen tension, local flow and carbon dioxide tension. In: Kessler M, Clark LC Jr, Lübbers DW, Silver IA, Simon W (eds) Ion and enzyme electrodes in biology and medicine. Urban and Schwarzenberg, München, pp 311–315

Lehmenkühler A (1979) Interrelationships between DC potentials, potassium activity, pO_2 and pCO_2 in the cerebral cortex of the rat. In: Speckmann E-J, Caspers H (eds) Origin of cerebral field potentials. Thieme, Stuttgart, pp 49–59

Lehmenkühler A (1982) Transient alkaline shift in the cortical tissue pH during the onset of spreading depression and of anoxic negative DC potential shift. Pflügers Arch 394:R49

Lehmenkühler A, Janus J (1978) Relation between local blood flow and extracellular potassium activity in the rat cerebral cortex during changes of pCO_2. Arzneimittel forsch 28:871

Lehmenkühler A, Bingmann D, Lange-Asschenfeldt H, Berges D (1978) Oxygen pressure and ictal activity in the cerebral cortex of artificially ventilated rats during exposure to oxygen high pressure. Adv Exp Med Biol 94:679–658

Lehmenkühler A, Lensing J, Caspers H, Janus J (1979) Biphasic reaction of extracellular Ca^{++} activity during interictal discharges in the brain cortex. Pflügers Arch 382:R42

Lehmenkühler A, Zidek W, Staschen M, Caspers H (1981) Cortical pH and pCa in relation to DC potential shifts during spreading depression and asphyxiation. In: Sykova E, Hnik P, Vyklicky L (eds) Ion-selective microelectrodes and their use in excitable tissues. Plenum, New York, pp 225–229

Lehmenkühler A, Zidek W, Caspers H (1982a) Changes of extracellular Na^+ and Cl^- activity in the brain cortex during seizure discharges. In: Klee MR, Lux HD, Speckmann E-J (eds) Physiology and pharmacology of epileptogenic phenomena. Raven, New York, pp 37–45

Lehmenkühler A, Staschen M, Caspers H (1982b) Depth profile of extracellular pH in the brain cortex during seizure activity. Pflügers Arch 394:R50

Lehmenkühler A, Caspers H, Kersting U (1984) Relations between DC potentials, ion activities and volume fraction in the extracellular space of the brain cortex with increasing pCO_2. Electronencephalogr Clin Neurophysiol 58:91P

Lehmenkühler A, Caspers H, Kersting U (1985) Relations between DC potentials, extracellular ion activites and extracellular volume fraction in the cerebral cortex with changes in pCO_2. In: Kessler M, Harrison DK, Höper J (eds) Ion measurements in physiology and medicine. Springer, Berlin Heidelberg New York Tokyo, pp 199–205

Lehmenkühler A, Kersting U, Richter A, Boerrigter P (1986) Relations of bioelectric activity, extracellular ion concentrations and extracellular volume in cortical epileptic foci. In: Speckmann E-J, Schulze H, Walden J (eds) Epilepsy and calcium. Urban and Schwarzenberg, Munch, pp 227–246

Li C-L, Jasper HH (1953) Microelectrode studies of the electrical activity of the cerebral cortex in the cat. J Physiol (Lond) 121:117–140

Libet B, Gerard RW (1941) Steady potential fields and neurone activity. J Neurophysiol 4:438–455

Libet B, Gerard RW (1962) An analysis of some correlates of steady potentials in mammalian cerebral cortex. Electroencephalogr Clin Neurophysiol 14: 445–452

Loeschcke HH (1956) Über den Einfluß von CO_2 auf die Bestandpotentiale der Hirnhäute. Pflügers Arch 262:532–536

Loeschcke HH (1971) DC potentials between CSF and blood. In: Siesjö BK, Sørensen SC (eds) Ion homeostasis of the brain. Munksgaard, Copenhagen, pp 77–96 (Alfred Benzon Symposium III)

Lux HD (1974) The kinetics of extracellular potassium: relation to epileptogenesis. Epilepsia 15:375–393

Lux HD, Heinemann U (1983) Consequences of calcium electrogenesis for the generation of paroxysmal depolarization shift. In: Speckmann E-J, Elger CE (eds) Epilepsy and motor system. Urban & Schwarzenberg, München, pp 100–119

Lux HD, Heinemann U, Dietzel I (1986) Ionic changes and alterations in the size of the extracellular space during epileptic activity. 44:619–639

Matsumoto H, Ajmone-Marsan C (1964a) Cortical cellular phenomena in experimental epilepsy. Interictal manifestations. Exp Neurol 9:286–304

Matsumoto H, Ajmone-Marsan C (1946) Cortical cellular phenomena in experimental epilepsy. Ictal manifestations. Exp Neurol 9:305–326

Matsumoto H, Ayala GF, Gumnit RJ (1969) Neuronal behavior and triggering mechanism in cortical epileptic focus. J Neurophysiol 32:688–703

Meyer JS, Denny-Brown D (1955) Studies of cerebral circulation in brain injury. I. Validity of continued local cerebral electropolarography, thermometry and steady potentials as an indicator of local circulatory and functional changes. Electroencephalogr Clin Neurophysiol 7:511–528

Monnier AM (1955) Die funktionelle Bedeutung der Dämpfung in der Nervenfaser. Ergeb Physiol 48:230–285

Moody W, Futamachi KJ, Prince DA (1974) Extracellular potassium activity during epileptogenesis. Exp Neurol 42:248–263

Morris ME (1974) Hypoxia and extracellular potassium activity in the guinea-pig cortex. Can J Physiol Pharmacol 52:872–882

Mutch WAC, Hansen AJ (1984) Extracellular pH changes during spreading depression and cerebral ischemia: mechanisms of brain pH regulation. J Cereb Blood Flow Metab 4:17–27

Mutch WAC, Hansen AJ (1985) Brain extracellular pH changes during alterations in substrate supply. In: Kessler M, Harrison DK, Höper J (eds) Ion measurements in physiology and medicine. Springer, Berlin Heidelberg New York Tokyo, pp 189–193

Neher E, Lux HD (1973) Rapid changes of potassium concentration at the outer surface of exposed single neurons during membrane current flow. J Gen Physiol 61:385–399

Nelson PG, Frank K (1963) Intracellularly recorded responses of nerve cells to oxygen deprivation. Am J Physiol 205:208–212

Nicholson C (1980) Modulation of extracellular calcium and its functional implications. Federation Proc 39:1519–1523

Nicholson C, Kraig RP (1981) The behavior of extracellular ions during spreading depression. In: Zeuthen T (ed) The application of ion-selective microelectrodes, Elsevier, Amsterdam, pp 217–238

Nicholson C, Phillips JM (1981) Ion diffusion modified by tortuosity and volume fraction in the extracellular microenvironment of the rat cerebellum. J Physiol (Lond) 321:225–257

Nicholson C, Kraig RP, Ferreira-Filho CR, Thompson P (1985) Hydrogen ion variations and their interpretation in the microenvironment of the vertebrate brain. In: Kessler M, Harrison DK, Höper J (eds) Ion measurements in physiology and medicine. Springer, Berlin Heidelberg New York Tokyo, pp 229–235

Niechaj A, van Harrefeld A (1968) Intracellular recording from cat's spinal interneurons during asphyxiation. Brain Res 8:54–64
O'Leary JL, Goldring S (1964) DC potentials of the brain. Physiol Rev 44:91–125
O'Leary JL, Goldring S (1976) Science and epilepsy. Neuroscience gains in epilepsy research. Raven, New York
Prince DA (1974) Neuronal correlates of epileptiform discharges and cortical DC potentials. In: Rémond A (ed) Handbook of electroencephalography and clinical neurophysiology, vol 2/C. Elsevier, Amsterdam, pp 2C/56–2C/70
Prince DA, Wilder BJ (1967) Control mechanism in cortical epileptogenic foci; "surround inhibition". Arch Neurol 16:194–202
Prince DA, Lux HD, Neher E (1973) Measurement of extracellular potassium activity in cat cortex. Brain Res 50:489–495
Prince DA, Pedley TA, Ransom BR (1978) Fluctuations in ion concentrations and seizures. In: Schoffeniels E, Franck G, Tower DB, Hertz L (eds) Dynamic properties of glia cells. Pergamon, Oxford, pp 281–303
Pumain R, Kurcewicz I, Louvel J (1983) Fast extracellular calcium transients: involvement in epileptic processes. Science 222:177–179
Rémond A (ed) (1974) Handbook of electroenphalography and clinical neurophysiology, vol 10/A. Elsevier, Amsterdam
Schneider MA Überlebens- und Wiederbelebungszeit von Gehirn, Herz, Leber, Niere nach Ischämie und Anoxie. Westdeutscher Verlag, Cologne
Siesjö K, von Hanwehr R, Norgelius G, Nevander G, Ingvar M (1985) Extra- and intracellular pH in brain during seizures and in the recovery period following the arrest of seizure activity. J Cereb Blood Flow Metab 5:47–57
Somjen GG (1975) Electrophysiology of neuroglia. Annu Rev Physiol 37:163–190
Somjen GG (1979) Extracellular potassium in the mammlian central nervous system. Annu Rev Physiol 41:159–177
Somjen GG (1980) Stimulus-evoked and seizure-related responses of extracellular calcium activity in spinal cord compared to those in the cerebral cortex. J Neurophysiol 44:617–632
Somjen GG (1984) Interstitial ion concentration and the role of neuroglia in seizures. In: Schwartzkroin PA, Wheal HV (eds) Electrophysiology of epilepsy. Academic, London, pp 302–341
Somjen GG, Aitken PG, Giacchino JL, McNamara JO (1986) Interstitial ion concentration and paroxysmal discharges in hippocampal formation and spinal cord. Neurol 44:663–680
Speckmann E-J, Caspers H (1966) Die sogenannte anoxische Terminaldepolarisation und ihre Beziehung zur Wiederbelebungszeit des Gehirns. Pflügers Arch 289: R2
Speckmann E-J, Caspers H (1967) Les modifications du potentiel continu cortical pendant l'arret respiratoire. Rev Neurol (Paris) 117:5–19
Speckmann E-J, Caspers H (1969) Verschiebungen des corticalen Bestandpotentials bei Veränderungen der Ventilationsgröße. Pflügers Arch 310:235–250
Speckmann E-J, Caspers H (1974) The effect of O_2 and CO_2 tensions in the nervous tissue on neuronal activity and DC potential. In: Rémond A (ed) Handbook of electroencephalography and clinical neurophysiology, vol 2/C. Elsevier, Amsterdam, pp 2C/71–2C/89
Speckmann E-J, Elger CE (1982) Neurophysiological basis of the EEG and of DC potentials. In: Niedermeyer E, Lopes da Silva F (eds) Electroencephalography. Basc principles, clinical applications and related fields. Urban and Schwarzenberg, München, pp 1–13
Speckmann E-J, Caspers H, Sokolov W (1970) Aktivitätsänderungen spinaler Neurone während und nach einer Asphyxie. Pflügers Arch 319:122–138
Speckmann E-J, Caspers H, Janzen RWC (1972) Relations between cortical DC shifts and membrane potential changes of cortical neurons associated with seizure activity. In: Petsche H, Brazier MAB (eds) Synchronization of EEG activity in epilepsies. Springer, New York, pp 93–111

Speckmann E-J, Caspers H, Janzen RWC (1978) Laminar distribution of cortical field potentials in relation to neuronal activities during seizure discharges. In: Brazier MAB, Petsche H (eds) Architectonics of the cerebral cortex. Raven, New York, pp 191–209

Staschen M, Zidek W, Lehmenkühler A, Caspers H (1981) Changes of extracellular ion activities (K^+, Na^+, Ca^{2+}, H^+, Cl^-) in relation to cortical DC potential shifts during reversible asphyxia. Pflügers Arch 389:R33

Sugaya E, Goldring S and O'Leary JL (1964) Intracellular potentials associated with direct cortical response and seizure discharge in cat. Electroencephalogr Clin Neurophysiol 17:661–669

Thomas RC (1976) The effect of carbon dioxide on the intracellular pH and buffering power of snail neurones. J Physiol (Lond) 255:715–735

Thomas RC, Meech RW (1982) Hydrogen ion currents and intracellular pH in depolarized voltage-clamped snail neurones. Nature 299:826

Tschirgi RD, Taylor JL (1958) Slowly changing bioelectric potentials associated with the blood-brain barrier. Am J Physiol 195:7–22

Urbanics R, Leninger-Follert E, Lübbers DW (1978) Time course of changes of extracellular H^+ and K^+ activities during and after direct electrical stimulation of the brain cortex. Pflügers Arch 378:47–53

Van Harreveld A, Stamm JS (1953a) Spreading cortical convulsions and depressions. J Neurophysiol 16:352–366

Van Harreveld A, Stamm JS (1953b) Cerebral asphyxiation and spreading cortical depression. Am J Physiol 173:171–175

Vyskocil F, Kriz A, Bureš J (1972) Potassium selective microelectrodes used for measuring the extracellular brain potassium during spreading depression and anoxic depolarization in rats. Brain Res 39:255–259

Washizu Y (1960) Effect of CO_2 and pH on the responses of spinal motoneurons. Brain Nerve 12:757–766

Withrow CD (1972) Systemic carbon dioxide derangement. In: Purpura DP, Kiffin Penry J, Tower DB, Woodbury DM, Walter (eds) Experimental models of epilepsy. Raven, New York, pp 477–494

Woody CD, Marshall WH, Besson JM, Thompson HK, Aleonard P, Albe-Fessard D (1970) Brain potential shift with respiratory acidosis in the cat and monkey. Am J Physiol 218:275–283

Wurtz RH, O'Flaherty JJ (1967) Physiological correlates of steady potential shifts during sleep and wakefulness. I. Sensitivity of the stady potential to alterations in carbon dioxide. Electroencephalogr Clin Neurophysiol 22:30–42

Zidek W, Lehmenkühler A, Caspers H, Lange-Asschenfeldt H (1978) Macromolecular buffering reverses the CO_2 effect on the membrane potential in snail neurons. Pflügers Arch 377:R43

Subject Index

GPSR Compliance

The European Union's (EU) General Product Safety Regulation (GPSR) is a set of rules that requires consumer products to be safe and our obligations to ensure this.

If you have any concerns about our products, you can contact us on ProductSafety@springernature.com

In case Publisher is established outside the EU, the EU authorized representative is:

Springer Nature Customer Service Center GmbH
Europaplatz 3
69115 Heidelberg, Germany

Batch number: 10620881

Printed by Printforce, the Netherlands